STATISTICAL CONCEPTS FOR ATTORNEYS

STATISTICAL CONCEPTS FOR ATTORNEYS

A REFERENCE GUIDE

Wayne C. Curtis

Q

QUORUM BOOKS
WESTPORT, CONNECTICUT
LONDON, ENGLAND

Library of Congress Cataloging in Publication Data

Curtis, Wayne C.
Statistical concepts for attorneys.

Bibliography: p.
Includes index.
1. Law—United States—Statistical methods.
2. Statistics. I. Title.
KF320.S73C87 1983 349.73'072 82-24068
ISBN 0-89930-033-2 (lib. bdg.) 347.30072

Library of Congress Catalog Card Number: 82-24068
ISBN: 0-89930-033-2

First published in 1983 by Quorum Books

Greenwood Press
A division of Congressional Information Service, Inc.
88 Post Road West, Westport, Connecticut 06881

Printed in the United States of America

10 9 8 7 6 5 4 3 2 1

Copyright Acknowledgment
Grateful acknowledgment is given for the permission to use excerpts from "Do Polls Tell the Story," a copyrighted interview in *U.S. News & World Report,* October 5, 1964. Used with the permission of *U.S. News & World Report.*

To the memory of my father

CONTENTS

FIGURES

TABLES

PREFACE

Societal changes during the twentieth century have been marked by an increasing utilization of statistics in nearly every field of endeavor. During the past two decades, statistics has become even more important in conceptualizing new relationships and in evaluating patterns of difference. Today such diverse fields as agriculture, natural science, social science, engineering, nursing, and psychology rely on statistics to express relationships and to evaluate products.

A growing proportion of legal issues involves the use of numerical data. Statistics, which deals with the analysis and interpretation of numerical data, is a useful tool to aid the courts in decision making. This book is a basic reference for attorneys who have limited or no prior knowledge of the field. In addition to legal practitioners in need of a basic reference, the user may also be a law school professor who wishes to review a specific concept or a student in the classroom who utilizes the book as a supplemental text. Nevertheless, the primary purpose is to introduce elementary statistics concepts to the practicing attorney; the focus of the material is on a broad range of legal applications. Attention is devoted to explaining how attorneys may encounter various statistical concepts—and how they may understand and effectively use them—in everyday situations. There is no attempt to provide an exhaustive discussion of each topic; rather the thrust is toward introducing and illustrating basic concepts. For those who wish to pursue additional reading on various topics, selected references are provided at the end of each chapter.

The material assumes no more than a working knowledge of basic algebra. Concepts and principles are developed without tedious and sophisticated derivations. Throughout the book the emphasis is on introducing topics on an intuitive level and then developing these topics into formal statistical concepts. Every attempt has been made to avoid purely academic presentations of the type found in college textbooks.

I believe attorneys will see inference as the objective of statistics. The average practicing attorney understands why decision procedures based on statistical inference function better than those based on intuition. Thus, an awareness of the role probability plays in drawing inferences is essential. Consequently, I have attempted to fuse the theories of probability and inference throughout the book.

In addition to the selected references noted earlier, each chapter contains actual cases and problems illustrating the specific legal application of the concept under discussion. To the extent possible, full citations of cases where certain statistical techniques have been applied are presented.

I am indebted to many persons for their interest and encouragement in the preparation of this manuscript. In particular, appreciation is expressed to my colleague, William H. Smith, statistician nonpareil, for his careful review and constructive criticism of the work; to Lowell E. Wilson, my coauthor in several publishing endeavors, for his invaluable assistance in providing reference materials for me; to my former colleague, Donald C. Hines, for graciously allowing me to draw upon some of his unpublished materials for Chapter 13; and especially to my wife, Seretha, for typing drafts of the manuscript as well as encouraging me to continue with the project. Of course, I assume full responsibility for any errors, omissions, or inconsistencies.

PART I

Statistical Description

CHAPTER 1

USES OF STATISTICS IN LITIGATION

> Suppose one takes a well-shuffled pack of cards, it is quite likely or not unlikely that the top card will prove to be a diamond: the odds are only three to one against; but most people would not say that it is quite likely to be the nine of diamonds for the odds are then fifty-one to one against. On the other hand I think that most people would say that there is a serious possibility or a real danger of its being turned up first and, of course, it is on the cards.
>
> —Lord Reid[1]

Contrary to Lord Reid's statement regarding probabilities, statistics is perhaps one of the most misunderstood and misused tools available to attorneys. In many instances, statistics and its applications are popularly misunderstood to include only collecting and analyzing numbers. Nothing could be further from reality. When statistical analyses are performed by those trained in the discipline, the methods are essentially those common to all fields of scientific empiricism.

STATISTICAL AWARENESS

The confusion among attorneys results primarily from the fact that *statistics* has two separate and distinct meanings. *Descriptive statistics* refers to collecting, organizing, and analyzing data. *Inferential statistics* is the drawing of conclusions from data that have been collected, organized, and analyzed. To illustrate this distinction, the arithmetic mean of a sample of observation is descriptive; if it is used to estimate the arithmetic mean of a larger group from which the sample was drawn, inferential techniques are involved.

Most attorneys will see inference as the objective of statistics. The foundation of statistical inference is probability theory. The object of statistical inference is to use information obtained from samples collected through the process of observation to draw conclusions about populations. Using probability theory to solve societal problems on a large-scale basis, however, is of relatively recent origin but is of increasing importance.

Lawyers, judges, and legal scholars can no longer ignore the field of statistics. One of the primary factors in this is the increasing intrusion of the federal government into the economy through various forms of regulation. Within recent years, courts have confronted larger and larger numbers of technical disputes involving problems such as air and water quality, the safety and economics of nuclear generators, the deregulation of natural gas, and the presence of carcinogenic agents in the workplace and in the kitchen.

A second factor compelling the use of statistics is the proliferation of data available to society. Examples are crime rates, health statistics, economic data, educational levels and achievements, and employment statistics. This growth in numbers has been accompanied by increased sophistication in the techniques used to analyze and interpret these numbers. Instead of simple tabular presentations, courts are confronted by multiple regression analyses, correlation coefficients, and tests of hypotheses.

Increased awareness on the part of attorneys poses questions as to the extent to which the law should employ statistics and the methods of inference that accompany them, what norms or guides should be established, and what aspects of statistics lawyers need to understand before they confront these pressing questions. Answers to the first two questions are beyond the scope of this work. The latter, which is the subject of this book, can perhaps be addressed initially by tracing the historical role of statistics in the litigation process, including recent innovative applications.

THE ROLE OF STATISTICS IN LITIGATION

One of the earliest reported instances of proof by statistical probabilities occurred in the infamous trial in 1899 of Alfred C. Dreyfus. To identify writing in a document as that of Dreyfus, the prosecution's witnesses reported a number of close matches between the lengths of certain words and letters in a document and the lengths of certain words and letters in correspondence taken from Dreyfus's home. As later noted by the panel of experts appointed to review the evidence in the *Dreyfus* case, there was nothing statistically remarkable about the existence of close matches in some word lengths between the disputed document and Dreyfus's correspondence. Thus, one of the first examples of introducing probabilities as proof ended on an unsatisfactory note.

The state of knowledge has advanced such that statistics are readily used in court cases today. Statistical data have been used extensively in civil rights litigation. For instance, statistics have been allowed as proof in jury discrimination,[2] school,[3] voting,[4] housing,[5] and age discrimination[6] cases. Reliance upon statistical evidence by the courts has been substantial in some instances. In *Alabama v. United States*, the court stated that "statistics often tell much and the courts listen."[7] In *Jones v. Lee Way Motor Freight, Inc.*, the court observed that "statistics often demonstrate more than the testimony of many witnesses."[8]

Antitrust cases are another area that has relied upon statistics to establish the relevant market as well as the defendant's share thereof. The courts generally view degree of concentration as a proxy for degree of competition in a particular market—an important element of many antitrust cases and of antitrust policy in general. In *United States v. United Shoe Machinery Corp.*, a case utilizing both descriptive and inferential statistics, the court concluded:

Through these leases United maintains a network of contracts with approximately 90 percent of all shoe factories. It supplies more than 75 percent of the demand for shoe machinery (excluding dry thread sewing machinery). From these and other facts, such as an analysis of United pricing practices, and a comparison between United and its competitors in terms of financial resources, of facilities, of accumulated experience, and of variety of machines offered, it appears that United has such substantial market power as to give it effective control of the shoe machinery market.[9]

Statistical presentations have been widely used in court cases involving deceptive advertising. In the landmark case *Bristol-Myers Co. v. FTC*, which involved inferential statistics based on sampling, the court held:

Based on the results of the survey, the corporation published advertisements from which the casual reader would reasonably infer that careful inquiry amongst the members of the dentist profession had disclosed that a large majority of the dentists in this country not only used Ipana but recommended it to their patients. . . . We are of the opinion that these sweeping statements were not justified by the answers to the questionnaire and that in consequence, as the Board found, the advertisements were misleading and likely to deceive the public.[10]

The most publicized use of statistical proof has occurred in Title VII cases. Courts generally have accorded statistical evidence great, if not decisive, weight. Dorsaneo asserts that the "relevance of statistical evidence is premised upon the legal principle that a statistical demonstration of the unequal consequences of a particular employment practice or selection process establishes a prima facie case of discrimination."[11] Limitations that may have restricted the courts in their enforcement of the act appear to have been eliminated by the Supreme Court in the landmark decision *Griggs v. Duke Power Co.* The court held that "congress directed the thrust of the Act to the *consequences* of employment practice, not simply the motivation."[12] This decision has generally resulted in the assumption that legally significant inferences can be obtained by comparing the composition of an employer's work force with that of the population from which the employees are drawn. For example, in *Hazlewood School District v. United States*, the Supreme Court held:

We noted that statistics can be an important source of proof . . . since absent explanation, it is ordinarily to be expected that nondiscriminatory hiring practices will in time result in a work force more or less representative of the racial and ethnic composition of the population from which employees are hired . . . where gross statistical disparities can be shown, they alone may in a proper case constitute prima facie proof.[13]

Like the courts, a number of regulatory agencies have embraced the use of statistics. Particularly appropriate for these agencies has been regression analysis, a concept by which the value of some dependent variable is estimated through use of one or more independent variables. The technique, long used by public and private decision makers in formulating policies, has gained acceptance during the past ten years in sharply contested regulatory proceedings. Regression studies have, for example, been used successfully in proceedings before the Commodity Exchange Authority,[14] Civil Aeronautics Board,[15] and Federal Communications Commission[16] as well as various state public utility commissions.

A NOTE TO THE READER

Problems posed in the courts result from conditions and needs of society. Because of technical and societal changes, a growing proportion of legal issues involves the use of numerical data. As has been noted, statistical analysis is a useful tool to aid the courts in decision making. Using samples drawn from a population or universe, statisticians employ a system of logical reasoning through which statistical inferences can be made and the risks of error associated with the inferences can be quantified.

This book has been written as a basic reference for attorneys who have limited or no prior knowledge of the field. The primary purpose is to introduce elementary statistics concepts to the practicing attorney. Throughout the book, the emphasis is on introducing topics on an intuitive level and then developing these topics into formal statistical concepts. Appropriate use of microcomputers, an innovation that has far-ranging legal application, is also woven into the discussion at key points in the text.

The book is organized into three sections. Part I, comprising Chapters 1 through 4, deals with statistical description. Presented in Chapter 2 are various means of organizing data. Histograms, frequency distributions, frequency polygons, and other graphical methods are discussed. Chapter 3 concentrates on measures of central tendency, i.e., the mean, median, and mode. Since measures of central tendency do not provide adequate information about a set of data, measures of dispersion are presented in Chapter 4. Specific attention is devoted to the range, the standard deviation, and the variance.

Part II, covering Chapters 5 through 8, serves as an introduction to statistical inference, a key concept for attorneys. Chapter 5 explains how the ideas of statistical inference rest upon basic concepts from probability theory. The binomial distribution, the first theoretical distribution to be discussed, appears in Chapter 6. Chapter 7 is devoted to a discussion of the normal distribution, the most important distribution in statistics. Statistical decision theory, or Bayesian analysis, is the subject of Chapter 8; the study of decisions under conditions of risk and uncertainty is discussed.

Chapters 9, 10, and 11 form the core of statistical inference. Chapter 9 presents the basis of sampling and how it relates to statistical inference. Examples are presented of the primary methods of sampling and the advantages and disadvantages of each. Presented in Chapter 10 is the concept of statistical estimation, including both point and interval estimates of population parameters at given confidence levels. The key element of statistical inference—hypothesis testing—is discussed in Chapter 11. Examples and problems presented in the chapter illustrate the widespread applicability of this important procedure.

The final three chapters concentrate on topics that are frequently of concern to practicing attorneys who specialize in certain types of cases. A nonparametric test, chi-square, is presented in Chapter 12. The basic use of this statistical test is to measure differences between observed and expected frequencies when nothing is known about the population from which a sample is taken. Correlation and regression, collectively known as association analysis, are the topic of Chapter 13. Correlation analysis measures the degree of association or relationship between sets of data, e.g., the relationship between cigarette smoking and the incidence of lung cancer, and is a widely used tool in many legal cases. Regression analysis, used to estimate the value of a dependent variable by knowing the value of one or more independent variables, is a well-known technique for forecasting statistical values. The last chapter of the book is concerned with index numbers. Index numbers are special ratios whose purpose usually is to show relative change over time. Most attorneys, particularly those involved in arbitration or wage disputes, are commonly exposed to such index calculations as the consumer price index and the wholesale price index.

Readers should not be alarmed at the mathematical notation and symbols. The material assumes no more than a working knowledge of basic algebra. Concepts and principles are developed without tedious or sophisticated derivations.

Finally, for those who wish to pursue additional reading on various topics, selected references are provided at the end of each chapter.

NOTES

1. The Heron II (Koufos v. C. Czarnikow, Ltd) 3 All E.R. 694-95 (H.L.).
2. Turner v. Fouche, 396 U.S. 346, 360 (1970).
3. See United States v. Hines County Board of Education, 417 F. 2d 852, 858 (5th Cir. 1969).
4. Alabama v. United States, 304 F. 2d 583, 586 (5th Cir.) *aff'd per curiam*, 371 U.S. 37 (1962).
5. See Kennedy Park Homes Ass'n v. Lackawanna, 436 F. 2d 108, 113 (2d Cir. 1970), *cert. denied*, 401 U.S. 1010 (1971).
6. See Hodgson v. First Fed. Sav. and Loan Ass'n, 455 F. 2d 818, 828 (5th Cir. 1972).
7. 304 F. 2d 586 (5th Cir.), *aff'd per curiam*, 371 U.S. 37 (1962).
8. 431 F. 2d 247 (10th Cir. 1970), *cert. denied*, 401 U.S. 954 (1971).
9. 110 F. Supp. 297-99 (D. Mass. 1953).
10. 185 F. 2d 60, 61 (4th Cir. 1950).

11. William V. Dorsaneo III, "Statistical Evidence in Employment Discrimination Litigation: Selection of the Available Population, Problems, and Proposals," *Southwestern Law Journal* 29 (1975): 859–60.

12. 401 U.S. 432 (1971).

13. 433 U.S. 307-08 (1977).

14. *In re* Cargill No. 120 (C.E.A., Aug. 13, 1970), *aff'd sub nom.* Carsill, Inc. v. Hardin, 452 F. 2d 1154 (8th Cir. 1971), *cert. denied,* 406 U.S. 932 (1971).

15. Domestic Passenger Fare Investigation Phase 7, 56 C.A.B. 211 (1971).

16. *In re* Western Union Tel. Co., No. 18935 (F.C.C., filed Apr. 12, 1970).

CHAPTER 2

DATA ORGANIZATION

Statistical data result from the process of observing, collecting, and recording various types of activities. Interpreting and using this data may, however, pose problems. In dealing with statistical data, the statistician first faces the problem of making sense of a large and confusing set of numbers, which are often presented in no logical order. This is accomplished through the methods of descriptive statistics.

Statistical procedures are designed to obtain usable information from a set of data. Without knowledge of the methods of data organization, one might find this an unsurmountable task. Through use of appropriate statistical techniques, large masses of data can be reduced to manageable proportions. Keep in mind, however, there is no statistical procedure that can extract more information from a set of data than was in the data to begin with.

FREQUENCY DISTRIBUTIONS

Suppose that demographic data were collected on the 50 employees in a large law firm in January 1983. To be more specific, each employee was asked to complete a questionnaire asking for his sex, age, marital status, number of children, and family income. The responses are listed for each employee along with his social security number. The personnel manager is now confronted with a list of 250 numbers, or 5 for each employee. Useful information is contained in the data. The arrangement of the numbers, however, is completely unrelated to magnitude. The personnel manager must find some means of extracting the useful information.

To illustrate how this may be done, assume the only data to be used from the survey is age of employee. Table 2.1 illustrates the age distribution of the 50 employees interviewed.

Before useful information can be obtained from this table, the data must be rearranged. This is done by arranging the numbers in order of size—either from largest to smallest or smallest to largest. Arrangement of the data is called a statistical array. Table 2.2 shows the data in an array.

Table 2.1
Tabulation of Ages of 50 Employees,
January 1983

43	41	39	37	42
55	40	45	51	58
31	50	50	66	25
36	61	43	30	47
41	39	28	38	27
39	40	26	46	39
58	51	29	42	55
40	41	49	34	53
62	54	49	57	32
46	56	60	64	29

Some of the characteristics of the data now become apparent. The range of values is from 25 to 66. The midpoint—the number located in the approximate center of the array—is 43. This could be considered a fairly representative age of all 50 employees.

Even an array is confusing when there are a great many numbers. It is usually desirable to condense the data into a more compact form that tends to highlight the more important characteristics. From a statistical point of view, the most important form of organizing data is called a *frequency distribution*. The data are organized into *classes* whose boundaries are given, e.g., 35 to 45 years of age. Each class has a *midpoint*, which is the average of the class boundaries, 40 in the above example. The *frequency* of each class is the number of observations between the boundaries of that class.

Two major problems arise in constructing frequency distributions. The first involves the number of classes into which the data are grouped. Determining the

Table 2.2
Tabulation of Ages of 50 Employees,
Arranged in Order of Magnitude,
January 1983

25	36	41	47	55
26	37	41	49	56
27	38	41	49	57
28	39	42	50	58
29	39	42	50	58
29	39	43	51	60
30	39	43	51	61
31	40	45	53	62
32	40	46	54	64
34	40	46	54	66

Table 2.3
Frequency Distribution of Ages of 50 Employees, January 1983

Age Group	Number of Employees
25 but under 35	10
35 but under 45	17
45 but under 55	13
55 but under 65	9
65 but under 75	1
Total	50

number of classes in the distribution is somewhat arbitrary. This decision depends primarily upon the judgment of the analyst, though certain guidelines do apply. Too few classes cause all of the distributional detail to be obscured, and too many classes result in unnecessary detail.

The second problem is that of defining the individual classes. Each class interval must have an upper and lower class limit; these specify the highest and lowest value included in that class. Class limits must be stated so that each data element belongs to one and only one class. Class limit specification depends on the nature of the data being grouped and the way the data are recorded (Table 2.3).

The frequency distribution in Table 2.3 is often called an *absolute frequency distribution* because it provides the absolute number of values that fall into each class. In some circumstances characteristics of the data can be emphasized by some other arrangement of the frequencies. The most common arrangement is called a *cumulative frequency distribution.* This distribution is constructed from the absolute distribution by starting at the lowest class and successively adding the frequencies as shown in Table 2.4, a cumulative distribution of ages of employees.

Table 2.4
Cumulative Frequency Distribution of Ages of 50 Employees, January 1983

Age	Cumulative Frequency
Less than 25	0
Less than 35	10
Less than 45	27
Less than 55	40
Less than 65	49
Less than 75	50

Table 2.5
Distribution of 50 Employees by Sex, January 1983

Sex	Number of Employees
Male	38
Female	12
Total	50

Note that in converting an absolute frequency distribution to a cumulative distribution, the method of defining the classes is changed. The cumulative frequency that corresponds to each class designation represents the total number of observations that fall in that class and all the classes that fall below it. Therefore, for a cumulative distribution, each class is defined by a single limit and a description as to how the frequencies relate to that limit. For example, Table 2.4 indicates that 27 of the 50 employees in the law firm are less than 45 years of age; 10 are less than 35 years old.

Frequency distributions are sometimes constructed to illustrate qualitative rather than quantitative data. When some observable characteristic is not inherently numerical, that characteristic is called an *attribute*. Such a characteristic can be observed only by its presence or its absence. A *categorical frequency distribution* results when classes in a frequency are specified in terms of attributes rather than numerical values. The number of observations that fall into each category comprise the class frequencies. Table 2.5, consisting of a classification by sex of the 50 employees interviewed earlier, is an example of a categorical frequency distribution.

HISTOGRAMS

To most people, a picture is easier to comprehend than a set of numbers in a table; a picture in conjunction with a table is even more useful. A *histogram* is the most common and useful graph or picture of a frequency distribution. A histogram is a vertical bar graph in which each bar represents a class of a frequency distribution, the width of each bar is the difference of the class boundaries, and the height of each bar is the class frequency.

A histogram corresponding to the frequency distribution in Table 2.3 appears as Figure 2.1.

OTHER GRAPHIC METHODS

Of the various charts and graphs for displaying statistical data, the most useful is the histogram. An alternate method for graphing a frequency distribution, however, is a *frequency polygon*. A frequency polygon is a broken-line graph

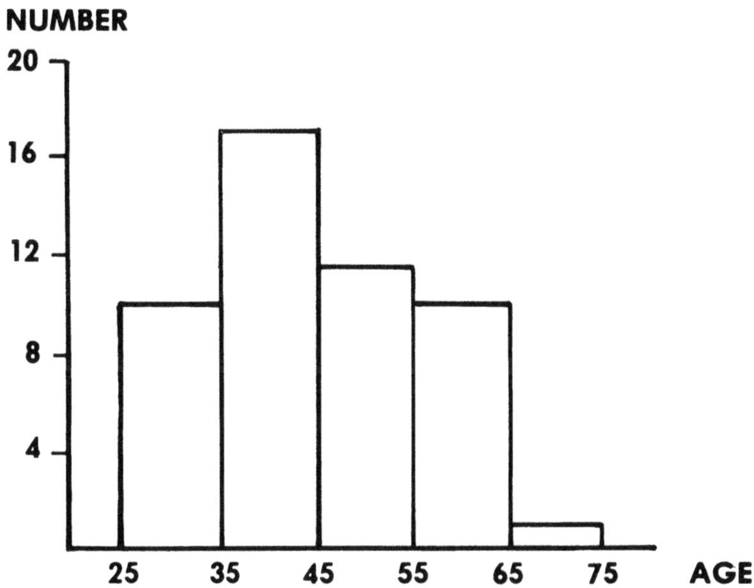

Figure 2.1
Histogram of Ages of 50 Employees

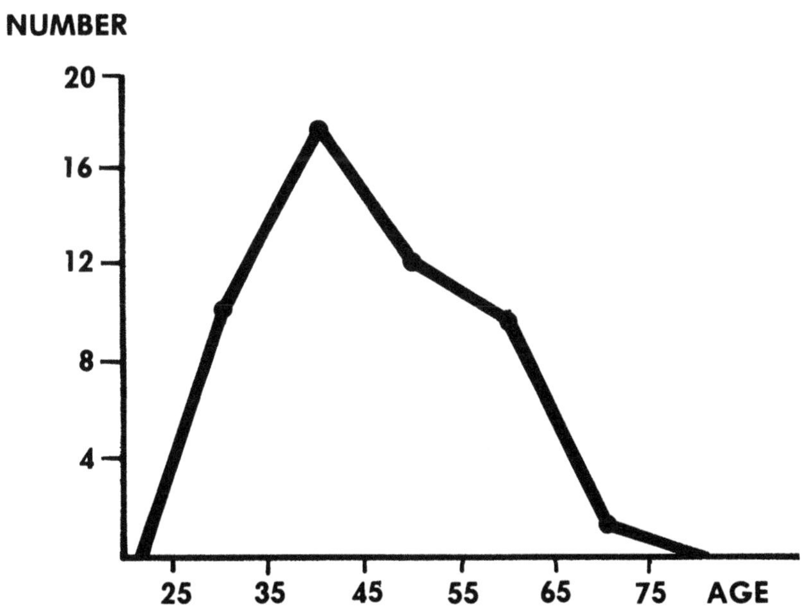

Figure 2.2
Frequency Polygon of Ages of 50 Employees

made by connecting the midpoints of the tops of the bars in a histogram. Classes with zero frequencies are added at both ends of the distribution so that the graph begins and ends on the horizontal axis. A frequency polygon of the data from Table 2.3 is shown in Figure 2.2.

For graphical presentations of categorical distributions, the *bar chart* is probably most commonly used. Similar to a histogram, the bars in the bar chart are used to represent the various classes. The height of each bar is proportional to the frequency of each class. The horizontal bar, however, is nonnumerical. A bar chart for the categorical distribution of Table 2.5 is shown in Figure 2.3.

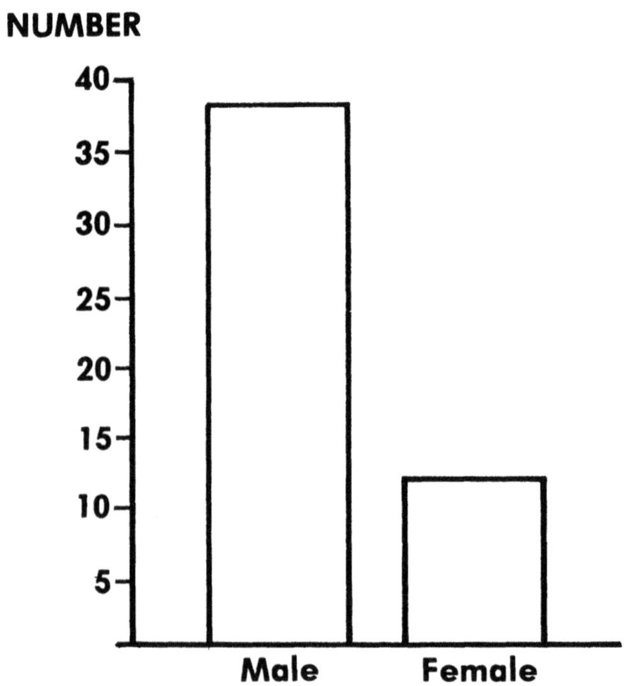

Figure 2.3
Bar Chart Showing Distribution of 50 Employees by Sex

GENERAL PURPOSE TABLES

When exact figures are necessary, a *general purpose* or *reference table* is used. Although frequency distributions or frequency tables need not follow any particular style other than the general rules that govern its construction, this is not the case with general purpose tables. These tables are published solely for the purpose of conveying certain types of statistical information; they are constructed for displaying statistical facts without having any special analytical purpose. Because of the exactness and precision required in a general purpose table, every general purpose table should consist of a minimum of four parts: title and table number, columns and stubs, source, and footnotes. These are illustrated in Table 2.6.

Every table should have a title indicating what the data are, how the data are classified, where the data originated, and what time period the data cover. If more than one table appears in a publication, each table should also have a number for identification.

The body of a table consists of a stub and one or more columns. The stub shows the classifications into which the data are grouped; the stub head identifies what is in the stubs. The column or columns contain the actual data, while the captions at the head of each column describe the contents of the column.

The source or origin of the data should always be indicated. Any explanatory information about the data should be provided in footnotes.

Table 2.6
Per Capita Tax Revenues, by Type of Tax, Selected Southern States, 1980

	Type of Tax (dollars)		
State	*Sales*	*Income*	*Total*
Alabama	294.67	110.95	405.62
Florida	363.83	—[a]	363.83
Georgia	275.79	159.60	435.39
Louisiana	283.13	58.86	341.73
Mississippi	356.91	59.62	416.53
North Carolina	240.56	200.97	441.53
South Carolina	301.15	158.64	459.79
Tennessee	308.42	6.71	315.13
Virginia	226.78	206.32	433.10

SOURCE: U.S. Bureau of the Census, *State Government Finances in 1980*, Washington, D.C., Government Printing Office, 1981, pp. 12–18.

[a]No state income tax.

ILLUSTRATIVE USES OF FREQUENCY DISTRIBUTIONS AND HISTOGRAMS

The primary purpose of constructing frequency distributions, histograms, and reference tables is to summarize large quantities of numerical data. Presentations using these forms make it much easier to comprehend the significant data characteristics and obtain useful information. The following examples illustrate how these techniques are used.

A statistician was recently employed to evaluate the method of charging rentals to lessees of the Fairhope Single Tax Corporation, an organization whose purpose is to establish and conduct a model community or colony free from all forms of private monopoly and to secure to its members equality of opportunity, the full reward of individual efforts, and the benefits of cooperation in matters of general concern. In a suit against the Fairhope Single Tax Corporation, several lessees alleged economic discrimination in charging of rentals to certain lessees, contending that rental charges were arbitrarily and subjectively charged.

The statistician employed by the plaintiffs in this case examined the pattern of rental charges imposed on all lessees by the Fairhope Single Tax Corporation during the period 1971–76. Data relating to changes in rental charges were collected through a sample of the master list of 1,196 lessees of the Fairhope Single Tax Corporation. The sample, consisting of 116 lessees, was randomly selected to ensure that all property sizes and locations were represented. After the initial sample was selected, it was disaggregated into two subsamples representing rural and urban lessees. Commonly accepted sampling procedures and statistical techniques were utilized throughout the study. That the sample selected was representative is illustrated by the fact that exactly 18 percent of those in the sample were rural lessees, which corresponds to the actual percentage as reported by the Fairhope Single Tax Corporation. Furthermore, 18.9 percent of the rental income in the sample was accounted for by rural lessees, as contrasted with 18 percent as reported by the corporation.

Aggregate changes in rental charges, reported on an annual percentage basis, are shown in Figure 2.4. As can be seen from Figure 2.4, the pattern of rental charges exhibited wide fluctuations. Rentals charged on all property increased at the rate of 15.2 percent per year. Urban property rentals increased 12 percent per year, while rural property rentals jumped 34.9 percent per year.

Although there were almost uniformly small increases in annual rental charges for urban property—about three-fourths of all urban lessees experienced increases of 10 percent or less per year—increases to rural property owners followed the pattern shown in Figure 2.5. Increases were concentrated in the ranges of 20–40 and 40–60 percent per year.

Apart and completely separate from the sample data—only a portion of which has been discussed here—the statistician was provided with documented evidence of increases in rents charged to various individual lessees. Even though the data in many cases covered longer time periods, they were, for purposes of the

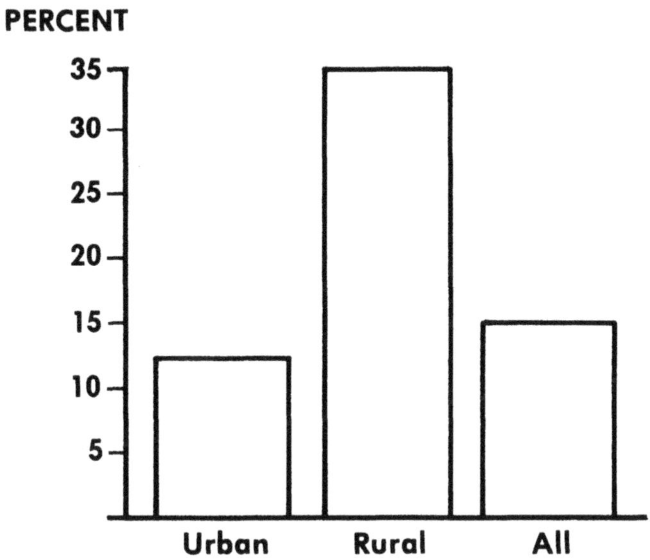

Figure 2.4
Annual Percentage Increases in Rental Charges, by Type of Property, Fairhope Single Tax Corporation

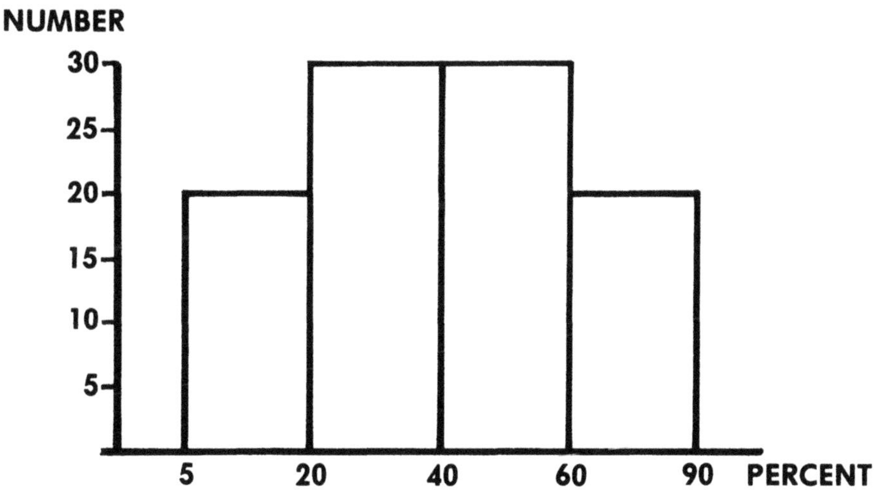

Figure 2.5
Histogram of Annual Percentage Increases in Rental Charges to Rural Lessees, Fairhope Single Tax Corporation

lawsuit, adjusted for the period 1971–76 to ensure comparability with the sample data. These data, which exhibit an exorbitant rate of increase in rentals, are shown in Figure 2.6.

The illustrations presented here, an abbreviated summary of the findings in the case, assisted in demonstrating a clear pattern of discrimination in rental charges to rural lessees.

A court case having several excellent illustrations of frequency distributions is *Pettway v. American Cast Iron Pipe Company.*[1] In this case, black employees of a manufacturer of cast iron pipes and ductile iron pipe and fittings brought a civil rights action for relief and damages, including back pay, for various allegedly discriminatory employment practices. The U.S. district court ordered and approved a plan for desegregating a governing board of the employer but, in a separate opinion, denied all other requested relief except attorney fees and costs. The employees appealed.

The court of appeals reversed and remanded the decision of the district court. Among other findings, the court held that testing and educational requirements

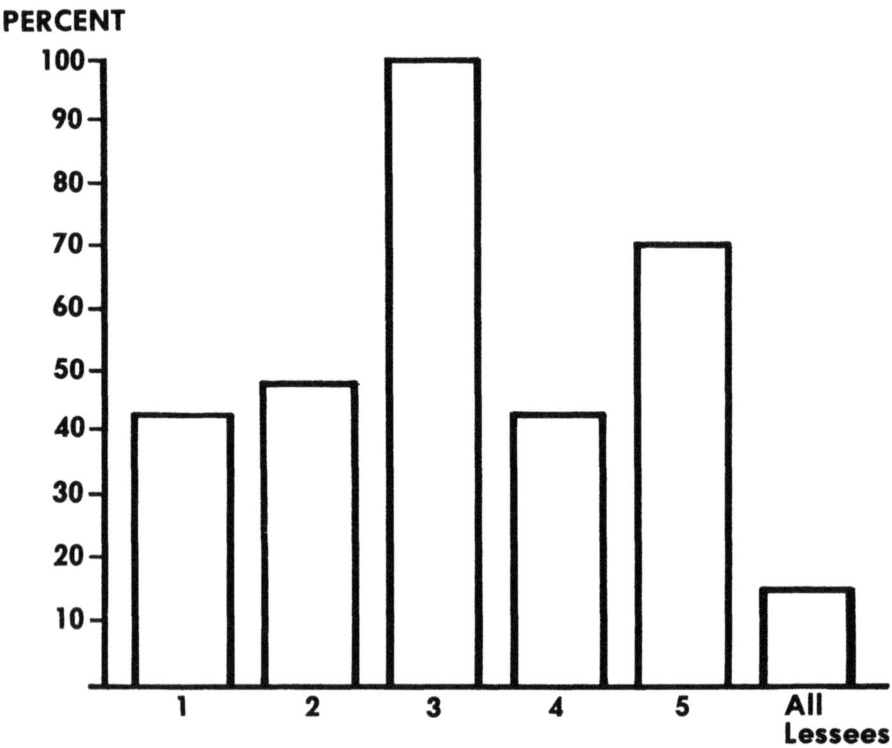

Figure 2.6
Annual Percentage Increase in Rentals to Five Lessees, Fairhope Single Tax Corporation

Table 2.7
Segregation by Types of Jobs

Department	Year	Number of Jobs			
		All Black	*All White*	*Black and White*	*Total*
Foundry	1963	38	34	0	72
	1965	37	30	6	73
	1969	18	29	21	68
	1971	19	24	22	65
Melting	1963	20	13	0	33
	1965	11	16	11	38
	1969	9	17	16	42
	1971	6	14	18	38
Steel Foundry	1963	18	19	0	37
	1965	14	22	4	40
	1969	9	23	11	43
	1971	9	20	12	41
Machine Shop	1963	5	19	0	24
	1965	—	—	—	—
	1969	5	11	4	20
	1971	5	26	4	36
Construction	1963	2	9	0	11
	1965	3	11	0	14
	1969	6	15	0	21
	1971	7	21	0	28
Shipping	1963	6	6	0	12
	1965	9	6	0	15
	1969	7	10	3	20
	1971	8	11	3	22
Personnel	1963	9	2	0	11
	1965	9	2	0	11
	1969	8	4	0	12
	1971	—	—	—	—
Technical	1963	4	9	0	13
	1965	3	10	0	13
	1969	3	3	0	6
	1971	1	1	0	2

SOURCE: Pettway v. American Cast Iron Pipe Company, 494 F. 2d 229 (1974).

NOTE: Dash indicates data not reported for these years.

which had been imposed by the employer in the past for hiring, promotion, and transfer and entry into the employer's apprentice and on-the-job training programs had been discriminatory relative to the employment opportunities of blacks. Further, the court concluded that the consequences of such past discrimination were currently being carried forward by otherwise neutral practices of departmental seniority and bidding and posting procedures. These practices collectively determined promotion, age requirements for entry into the company's apprentice programs, and departmental bidding utilized for selection of on-the-job trainees.

Although there are numerous charts and tables reported in this case, chart D—which is reprinted here as Table 2.7—is perhaps most illustrative. Addressing the question of segregation by types of jobs, the court held:

> The chart drawn up by the court reflects that in 1963 not a single type of job in any of the eight key departments was held by *both* white and black employees; 213 positions were totally segregated. The data for 1969 indicate that there were 55 mixed positions and 177 all white or all black positions. But the pertinent indicium for the period 1965–1969 is that the previous *all black* positions were being integrated by the influx of white personnel rather than any upward movement by black employees. In 1965 there were 86 all black jobs; in 1969 these were reduced to 60 jobs. A total of 34 positions were integrated during this period. In other words, 26 of the 34 jobs were integrated by the movement of *white* employees or applicants into previously *all black* jobs. As late as 1971, only 59 of the 232 jobs were integrated—only 25 percent of the total.[2]

NOTES

1. 494 F. 2d 211 (1974).
2. *Id* at 229–30.

SELECTED REFERENCES

Blalock, Hubert M., Jr. *Social Statistics,* Chap. 5. New York: McGraw-Hill Book Co., 1960.

Campbell, Stephen K. *Flaws and Fallacies in Statistical Thinking,* Chap. 5. Englewood Cliffs, N.J.: Prentice-Hall, 1974.

Huntsberger, David V. *Elements of Statistical Inference,* Chap. 2. 2d ed. Boston: Allyn & Bacon, 1967.

Pfangel, J. *Theory of Measurement.* New York: John Wiley & Sons, 1968.

CHAPTER 3

MEASURES OF CENTRAL TENDENCY

Various procedures for organizing statistical data were discussed in Chapter 2. These included grouping of data into several types of frequency distributions and constructing histograms, frequency polygons, and bar charts. The purpose behind these groupings is to arrange data to give a more accessible and accurate overview of comparisons and pertinent facts. This aspect of describing data is very useful. Often, though, one does not want to look at the entire set of data. Certain measures can be computed from the data which will summarize and/or describe some of the important characteristics of the data.

In this chapter, three measures of central tendency—the *mode*, the *median*, and the *mean*—will be considered. All of these are central numbers or averages about which data are arranged.

THE MODE

The mode is the simplest measure of central tendency. It is merely the value that occurs most often. A set of data may have a single mode, in which it is said to be *unimodal;* it may have two modes, which makes it *bimodal;* or it may have several modes and be called *multimodal*. On the other hand, for a given set of data, there may be no mode.

Look at the statistical array in Table 3.1, which summarizes the survey reported in Chapter 2. The mode, which occurs four times in the array, is 39. If the personnel manager wants to select a typical age among employees in the law firm, he might select 39, the modal age of those in the survey.

In a frequency distribution, it is not possible to find the exact value of the mode. Although there are several means of finding the value of the mode from grouped data, the class having the greatest frequency is designated the *modal class*. The mode of a frequency distribution is the midpoint of the modal class. The midpoint of the modal class is generally referred to as the *crude mode*. Since precise methods are seldom required, this method of specifying the mode is usually sufficient.

The frequency distribution for ages of employees is presented in Table 3.2.

Table 3.1
Ages of 50 Employees, Arrayed in Order of Magnitude, January 1983

25	36	41	47	55
26	37	41	49	56
27	38	41	49	57
28	39	42	50	58
29	39	42	50	58
29	39	43	51	60
30	39	43	51	61
31	40	45	53	62
32	40	46	54	64
34	40	46	54	66

The modal class is the class with limits of 35 and 45. The crude mode of this distribution is 40.

THE MEDIAN

A more useful and more common measure of central tendency is the median. The median is very simple in nature: it is merely the halfway point of the data. In other words, one-half of the observations will be less than the median and one-half will be greater. For example, the median family income in the United States in 1979 was $16,553. This means that half the families in the United States had an income greater than $16,553, while half earned less.

For ungrouped data arranged in the form of an array, the median is equal to the value of the middle element if *N*, the number of elements, is an odd number. If *N* is an equal number, the median is the value halfway between the two middle elements. For example, in the array in Table 2.1, there are 50 elements. There-

Table 3.2
Frequency Distribution of Ages of 50 Employees, January 1983

Age Group	Number of Employees
25 but under 35	10
35 but under 45	17
45 but under 55	13
55 but under 65	9
65 but under 75	1
Total	50

fore, the median value is the value halfway between the 25th and 26th elements—in this case, the value is 42.

There is a simple formula for locating the position of a median for ungrouped data. Md represents the value of the median, and Md_p represents the position of the median. If N represents the number of elements in an array, then

$$(3.1) \qquad Md_p = \frac{N + 1}{2}$$

For example, using this formula, the median of 3, 17, 6, 14, 11, 9, 21, 7 is 10. When arrayed in ascending order, these values become 3, 6, 7, 9, 11, 14, 17, 21. Since there are eight numbers, $Md_p = (8 + 1)/2 = 4.5$. The position of the median is 4.5; this means the median is that value located halfway between the fourth and fifth values from either end of the array. The average of the two middle elements is 10, i.e., $(9 + 11)/2$.

Using the same formula, the median value of 2, 2, 5, 7, 9, 9, 9, 10, 10, 11, 12, 18, 18, 19, 21 is 10. That is, $Md_p = (15 + 1)/2 = 8$. The eighth element in the array is 10.

In a frequency distribution of grouped data, the position of the median must be approximated. Since it is not possible to find a middle element, the first step is to find the *median* class. This class contains the middle or $N/2$ item, since by common practice it is specified that the median corresponds to the $N/2$ item. This is done by finding the cumulative frequency of each class until that class is found for which the cumulative frequency is greater than or equal to $N/2$. The value of the median within that class is found through the process of interpolation. The procedure is summarized by the following formula:

$$(3.2) \qquad Md = L + C\frac{(N/2 - F)}{f}$$

where

Md = value of the median
L = lower limit of the median class
C = class interval of the median class
N = number of observations or data elements in the set
F = cumulative frequency up to the median class
f = frequency of the median class

Take as an example the survey discussed earlier. One could determine the median age of employees by using the frequency distribution below.

Age	*Frequency (f)*	*Cumulative Frequency (f)*
25 but under 35	10	10
35 but under 45	17	27

Since the $N/2$ element has been specified as the median value, the 25th element in the set contains the median. The first class contains 10 elements and the second class has 17 elements; the cumulative frequency of the second class is 27. Therefore, element number 25, the median, must fall in the second class. Once the median class has been located, it is not necessary to accumulate the frequencies. The median class has the limits 35 and 45.

The approximate value of the median can now be determined by using Equation 3.2. L, the lower limit of the median class, is 35. The class interval of the median class is 10. F, the cumulative frequency up to the median class, is 10. Thus, the median (Equation 3.2) is

$$\begin{aligned}\text{Md} &= 35 + 10\,\frac{(50/2 - 10)}{27} \\ &= 35 + 10\,\frac{15}{27} \\ &= 35 + 5.6 = 40.6\end{aligned}$$

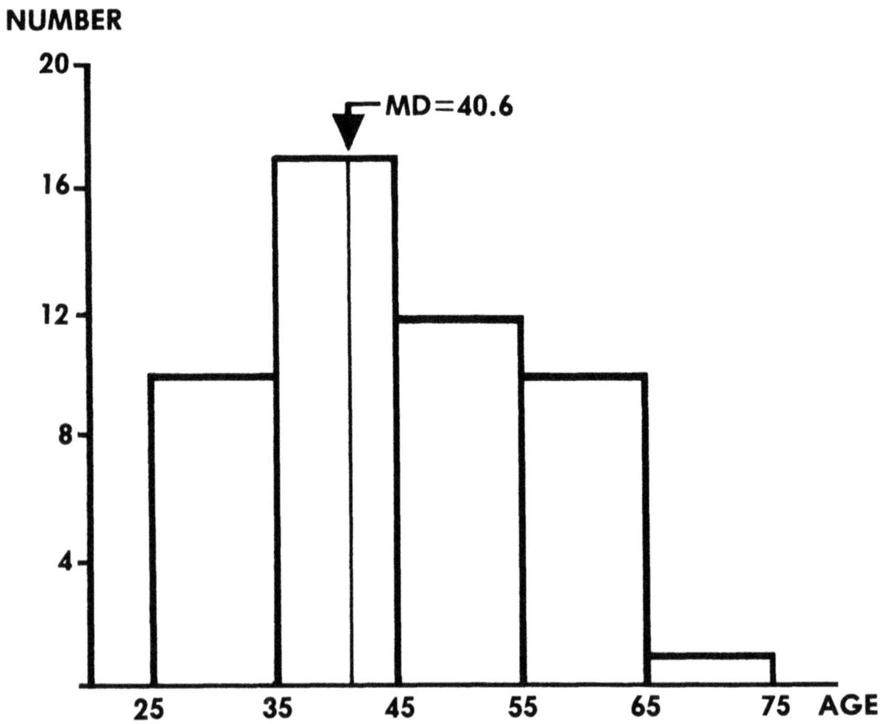

Figure 3.1
Histogram of Ages of 50 Employees Showing Median

The median of this frequency distribution is illustrated in Figure 3.1.

Since the median age is 40.6, the personnel manager knows that half the employees are less than 40.6 years of age and half are greater than 40.6. It was previously determined that the mode of this frequency distribution is 40.

THE ARITHMETIC MEAN

Because it is easily understood, best known, and most useful, the arithmetic mean is the most often used average in statistics. The arithmetic mean of a group of data can be thought of as a balance point. If the data are arranged on a horizontal scale, the mean would be the point at which the numbers on the left "balance" the numbers on the right. The term *mean,* when used without qualifying adjective, is the arithmetic mean.

The arithmetic mean can be defined as the sum of the values of a variable divided by the number of values in a set. To compute the mean of the values of a variable x_i, such as $x_1, x_2, \ldots, x_n$, the following equation can be used:

$$\mu = \frac{\sum_{i=1}^{N} x_i}{N} \tag{3.3}$$

where

μ = the arithmetic mean
x_i = the individual observations of the variable x
N = the number of observations
$\sum_{i=1}^{N}$ = the summation of the individual values of the variable x

When there is no confusion about the limits of the summation, the equation becomes

$$\mu = \frac{\sum x}{N} \tag{3.4}$$

If a set of data consists of the five values $x_1 = 15$, $x_2 = 20$, $x_3 = 11$, $x_4 = 6$, $x_5 = 8$, the mean of this data set (Equation 3.3) is

$$\mu = \frac{\sum_{i=1}^{5} x_i}{N} = \frac{60}{5} = 12$$

In a like manner, using Equation 3.3, the mean age of the employees in the law firm reported in Table 2.1 is 44.28 years.

Properties of the Mean

The mean is representative of a set of numerical values, in the sense that it represents a point of balance for the set. Several properties or characteristics of the arithmetic mean should be kept in mind. First, the arithmetic mean has the one weakness of being unduly influenced by unusually large or unusually small values in a distribution. For example, if five people earn $10,000, $12,000, $11,000, $14,000, and $95,000, the mean earnings are $28,400. This figure is not truly representative of the observations because of the undue influence of one unusually large value. Consider the following five earnings figures: $15,000, $18,000, $16,500, $19,000, and $17,000. The mean earnings here are $17,100—a much more representative figure than $28,400 in the previous example.

The arithmetic mean has another familiar property that will be useful to remember. The sum of the deviations of the individual values from their mean is zero. If the arithmetic mean is subtracted from every member of the set and the resulting differences added algebraically, the sum is zero. This can be illustrated by using the following set of five numbers, an example used previously where μ was found to be 12.

x	$x - \mu$
15	3
20	8
11	−1
6	−6
8	−4
	$(x - \mu) = 0$

The arithmetic mean has two other interesting and useful characteristics. First, if a constant is added to or subtracted from each value in a set of data, the mean of the resulting set is equal to the mean of the original set plus or minus the constant.

Set 1(x)	*Constant (c)*	*Set 2 (x + c)*
15	5	20
20	5	25
11	5	16
6	5	11
8	5	13
$\mu_1 = 12$		$\mu_2 = 17$

The mean of Set 2 is $\mu_1 + 5$.

The last property of the mean concerns multiplication of a set of data by a constant. If every value in a set of data is multiplied by a constant c, the mean of the resulting set is equal to the mean of the original set multiplied by the constant.

Set 1(x)	*Constant (c)*	*Set 2 (xc)*
15	5	75
20	5	100
11	5	55
6	5	30
8	5	40
$\mu_1 = 12$		$\mu_2 = 60$

The mean of Set 2 is the original mean multiplied by 5.

Weighted Arithmetic Mean

Equation 3.3 assumes that the values being averaged are all of equal importance. This, however, is not always the case. There are situations in which values in a data set do not have equal importance. In this case, computing the mean by the above equation may result in a value that is not truly representative and may, in fact, be misleading. For example, all the components included in the consumer price index are not of equal importance; the index is an example of a weighted average. To compute the mean of the values of the variable x when each has an assigned weight such as $w_1x_1, w_2x_2, \ldots, w_nx_n$,

$$\mu = \frac{\sum_{i=1}^{N} w_i x_i}{\sum_{i=1}^{N} w_i} \tag{3.5}$$

where

w_i = the weight assigned to the ith value in the set.

Assume, for purposes of determining a weighted average price of gasoline at ten service stations, the gasoline prices and weights reported in Table 3.3 are used. Weights are based on volume of sales at the stations. Using Equation 3.5,

$$\mu = \frac{\sum_{i=1}^{N} w_i x_i}{\sum_{i=1}^{N} w_i} = \frac{132.15}{100} = 1.32$$

Without weighting, the average price of gasoline using Equation 3.3 is \$1.37. The weighted value of \$1.32 reflects volume of sales and is considered the correct value.

In order to simplify the arithmetic involved, it is customary to assign weights so that they total either 1 or 100. However, this is not an absolute requirement.

Table 3.3
Computation of Weighted Arithmetic Mean of Gasoline Prices at Ten Locations, January 1983

Location	Price x (dollars)	Weight w (percent)	Weighted Arithmetic Mean wx
Station 1	1.50	5	7.50
Station 2	1.28	15	19.20
Station 3	1.44	5	7.20
Station 4	1.27	15	19.05
Station 5	1.31	15	19.65
Station 6	1.21	15	18.15
Station 7	1.49	5	7.45
Station 8	1.52	5	7.60
Station 9	1.29	15	19.35
Station 10	1.40	5	7.00
Total	13.71	100	132.15

Mean for Grouped Data

Equation 3.3 must be modified when data are grouped into a frequency distribution. Thus, the values of the individual observations are no longer available from the distribution. The frequency distribution specifies class intervals and the number of observations that fall into each class interval. To calculate the mean of the frequency distribution, it is first necessary to designate some value in each of the classes to be representative of all the observations in that class. This value is the midpoint of the class.

As an example, consider the frequency distribution of ages of employees first reported in Chapter 2. The mean computed from a frequency distribution is

$$\mu = \frac{\sum_{i=1}^{K} x_i f_i}{N} \tag{3.6}$$

where

x_i = midpoint of ith class
K = number of classes in the distribution
f_i = frequency of ith class
N = number of observations

Thus, the mean age of employees calculated from the frequency distribution is

$$\mu = \frac{2250}{50} = 45$$

Table 3.4
Frequency Distribution of Ages of 50 Employees, January 1983

Age Group	Midpoint (x)	Number of Employees (f)	Frequency Distribution xf
25 but under 35	30	10	300
35 but under 45	40	17	680
45 but under 55	50	12	600
55 but under 65	60	10	600
65 but under 75	70	1	70
Total		50	2,250

Remember the mean age calculated from ungrouped data was 44.28 (Equation 3.3). This is all shown in Table 3.4.

SUMMARY OF THE MEAN, THE MEDIAN, AND THE MODE

The three measures of central tendency discussed thus far are considered representative values of a set of data. The mean is by far the most commonly used of these averages. It uses every value in the data set; it also lends itself to algebraic manipulation. The primary disadvantage is that it is affected by extreme values in the data set. This is a particularly acute deficiency if the extreme values all fall on the same side of the data set. In this case the mean is less representative of the total set than the other averages.

Unlike the mean, a computational average, the median, a useful average, is positional in nature. The true middle point in the set, it divides the data in half. Even though the median does not lend itself to algebraic manipulation, the presence of extreme values does not have as great an effect on the median as on the mean. Therefore, if a data set contains extreme values, the median may be the most representative value.

The mode is the least useful of the three averages. It is most typical because it occurs most often, and it is unaffected by extreme values. However, the mode does not always exist in a set of data. In other instances, several modes may occur. If data are grouped in a frequency distribution, the mode may be difficult to determine.

ILLUSTRATIVE USES OF THE ARITHMETIC MEAN

In *L. C. L. Theatres, Inc. v. Columbia Pictures Industries,*[1] a motion picture exhibitor brought action against ten motion picture distributors and their law firm, alleging violations of the Sherman Act. The distributors counterclaimed to recover film rental from the exhibitor and its principal stockholder. The district

Table 3.5
Racial Distribution and Salary of Employees by Department, as of August 12, 1971

		Blacks				Whites		
Department	**Average Wage in Department**	*Number*	*Percentage*	*Average Wage (dollars)*	*Accumulating Percentage in Work Force*	*Number*	*Average Wage (dollars)*	*Accumulating Percentage in Work Force*
Inspection	$3.83	1	2.9	2.78	0.1	34	3.86	2.1
Maintenance	3.79	9	8.8	3.29	1.1	93	3.84	7.8
Machine shop	3.65	50	11.7	3.08	6.5	379	3.72	31.1
Electrical	3.63	4	5.1	2.98	6.9	74	3.67	35.6
Steel pipe	3.63	1	2.9	3.23	7.0	33	3.64	37.6
General stores	3.57	2	8.7	3.38	7.2	21	3.59	38.9
Construction	3.43	27	31.8	3.01	10.1	58	3.63	42.5
General yards	3.38	19	47.5	3.22	12.2	21	3.52	43.8
Shipping	3.36	30	46.9	3.06	15.5	34	3.63	45.9
Steel foundry	3.32	36	39.6	3.16	19.4	55	3.43	49.3
Foundry	3.28	247	44.7	3.03	46.2	305	3.48	68.0
Melting	3.27	75	39.9	3.14	54.4	113	3.35	74.9
Monocast	3.16	371	54.2	3.08	94.7	314	3.26	94.2
Bolt	3.11	20	34.5	2.91	96.9	38	3.22	96.5
Personnel	3.04	13	38.1	2.95	98.4	23	3.09	97.9
Total		905				1,595		

SOURCE: Pettway v. American Cast Iron Pipe Company, 494 F. 2d 230 (1974).

NOTE: All of ACIPCO's departments are included in this chart *except* those departments having fewer than 15 employees: Industrial Engineering, Production Control, Medical, Research, Laboratory, Purchasing, Finance, and Engineering. In total, 33 whites and 15 blacks are employed in these departments.

court held that the evidence established that the exhibitor engaged in fraud and breached his license agreements with the distributors by underreporting box office receipts and, among other findings, that damages to the motion picture distributors were $429,200. The last finding was based primarily upon the testimony of a statistician from the University of Texas, who served as an expert witness for the distributors.

A series of statistical tests was performed to determine if underreporting of income had occurred on the part of L. C. L. Theatres, which consisted of seven theaters. These calculations were based primarily upon arithmetic means. The court held:

Dr. Peterson then calculated damages. This was done by multiplying Dr. Peterson's calculated revenue for each of the seven theaters by the average film rental that it had paid. For example, the Surf Drive-In had revenue of $369,000.00, with average film terms of 33.4 percent, resulting in a film rental due of $123,400.00, less the rental paid of $78,600.00, equaling a sum due of $44,800. Following this process for the other six theaters, Dr. Peterson arrived at a damage figure of $429,200.00. . . . Because this figure is the product of negotiation between the distributor and the exhibitor, based on actual performance of the film, any underreporting would have had a deflating effect upon the final terms. . . . Dr. Peterson calculated a damage figure which employed contract terms and net final negotiated terms for film rental. . . . The damage figure, using the higher rental rate, is $776,100.00. Finally, Dr. Peterson took the mean of the two figures as his final damage sum—$602,600.00.[2]

In presenting his testimony, Dr. Peterson developed a series of schedules which, in the interest of brevity, are not reproduced here. In finding for the counterclaimant in the amount of $429,200, the court stated that the "attached schedules (prepared by the distributors) have been reviewed by the court. They reflect a proper breakdown among the distributors of the total damages found."[3] The court had earlier said that "[t]his conclusion is buttressed not only by the internal consistency of the calculations of the distributors, but by the court's confidence in the competency of Dr. Peterson."[4]

Another case that placed reliance on use of the arithmetic mean is *Pettway v. American Cast Iron Pipe Company*,[5] cited earlier in Chapter 2. In chart E, reproduced here as Table 3.5, weighted averages were used to show that blacks were paid lower wages than whites by the defendant. The court stated that "[t]his finding coupled with plaintiff's chart comparing the average wage of blacks and average wage of whites in all the departments firmly established [*sic*] that black workers were forced into lower paying positions within the company."[6]

NOTES

1. 421 F. Supp. 1099 (1976).
2. *Id* at 1098.

3. *Id* at 1103.
4. *Id* at 1102.
5. 494 F.2d 211 (1974).
6. *Id* at 230.

SELECTED REFERENCES

Blalock, Hubert M., Jr. *Social Statistics,* Chap. 5. New York: McGraw-Hill Book Co., 1960.

Boot, John C. G. and Edwin B. Cox. *Statistical Analysis for Managerial Decisions,* Chap. 5. New York: McGraw-Hill Book Co., 1974.

Chou, Ya-lun. *Statistical Analysis with Business and Economic Applications,* Chaps. 2–3. New York: Holt, Rinehart and Winston, 1975.

Guenther, William C. *Concepts of Statistical Inference,* Chap. 3. 2d ed. New York: McGraw-Hill Book Co., 1973.

Huntsberger, David V. *Elements of Statistical Inference,* Chap. 3. 2d ed. Boston: Allyn & Bacon, 1967.

Yamane, Taro. *Statistics: An Introductory Analysis,* Chap. 3. 3d ed. New York: Harper & Row, 1973.

CHAPTER 4

MEASURES OF DISPERSION

Chapter 3 discussed several measures of central tendency. To the extent that any single value can be representative, these measures are considered to be representative of a population. It should be kept in mind, however, that it is often misleading to use any one single measure, e.g., a mean, median, or mode, to describe a population. A single average tells nothing about *dispersion* or scatter within a distribution. Several values, in fact, are required to provide a reasonable description of a distribution.

To illustrate the inadequacy of any single average, consider two recently concluded law cases. The same group of six attorneys worked on both cases. Case A required a total of 100 billable hours; case B required 110 hours (Table 4.1).

Looking only at the measures of central tendency, one would conclude that the two cases are identical. The mode, median, and mean are 60 in each case. Consequently, in attempting to describe these two cases by using any of the three averages, one would have to conclude that they are identical. The data make it obvious, however, that they are not identical. Variation appears to be the principal difference. To provide an adequate description of the cases, some measure of variation is required in conjunction with measures of central tendency discussed in Chapter 3.

RANGE

The *range* is the simplest of all measures of dispersion or variation. It can be defined as the difference between the largest and smallest values of some variable. Or it may be defined as the largest and smallest values themselves.

For example, in case A the range is defined as 80, or 20 to 100. The latter expression, however, provides more information and is generally superior to the former. For case B, the range in hours is 10 to 110.

An advantage of the range is its ease of computation, but it tells nothing about what lies between the largest and smallest values. That is, its chief disadvantage is that it uses only part of the available data. The range is computed from only

Table 4.1
Billable Hours and Mode, Median, and Mean Hours for Two Cases, by Attorney, January 1983

	Hours	
Attorney	*Case A*	*Case B*
1	20	10
2	40	30
3	60	60
4	60	60
5	80	90
6	100	110
Mode	60	60
Median	60	60
Mean	60	60

two values in the set of data. It ignores the way the data vary for the remaining observations.

VARIANCE AND STANDARD DEVIATION

The most useful measures of dispersion and those with the most desirable mathematical properties are the variance and standard deviation. *Variance* may be defined as the arithmetic mean of the squared deviations of the individual items about their mean. The square root of the variance is called the *standard deviation.* The measure is particularly useful when dealing with the normal distribution (Chapter 7).

Variance and standard deviation possess several characteristics that cause these measures to be widely used:

1. They utilize all the data, i.e., every value in a distribution enters into the calculation of these measures.
2. The magnitudes of the measures are related to the variation present in the data, i.e., the value of the measure is large when variation is large and small when variation is small.
3. They measure the variation about the mean of the data set.

It should be noted that the standard deviation, denoted as σ, is always expressed in the same units as the original data, i.e., the standard deviation of the average wages received by production employees is expressed in dollars and cents. This is not true of the variance, σ^2, which is expressed in terms of the original units squared.

The definitional formulas for the variance and standard deviation of a population are as follows:

$$(4.1) \qquad \sigma^2 = \frac{\sum_{i=1}^{N} (x_i - \mu)^2}{N}$$

$$(4.2) \qquad \sigma = \sqrt{\frac{\sum_{i=1}^{N} (x_i - \mu)^2}{N}}$$

where

x_i = value of the *i*th observation in a set of data
μ = mean of the set
N = number of elements or observations in the set

Algebraically equivalent expressions of Equations 4.1 and 4.2 that are often easier to use in performing calculations are:

$$(4.3) \qquad \sigma^2 = \frac{\sum_{i=1}^{N} x_i^2}{N} - \left(\frac{\sum_{i=1}^{N} x_i}{N}\right)^2$$

and

$$(4.4) \qquad \sigma = \sqrt{\frac{\sum_{i=1}^{N} x_i^2}{N} - \left(\frac{\sum_{i=1}^{N} x_i}{N}\right)^2}$$

Note that $\Sigma x_i/N$ is simply the mean of the data set.

Both sets of formulas are used in Table 4.2 to calculate the variance and standard deviation for hours worked by attorneys on case A. Using Equations 4.1 and 4.2:

$$\sigma^2 = \frac{\sum_{i=1}^{N} (x_i - \mu)^2}{N} = \frac{4{,}000}{6} = 666.67$$

$$\sigma = \sqrt{666.67} = 25.82$$

Now, using Equations 4.3 and 4.4:

$$\sigma^2 = \frac{\sum_{i=1}^{N} x_i^2}{N} - \left(\frac{\sum_{i=1}^{N} x_i}{N}\right)^2 = \frac{25{,}600}{6} - (60)^2 = 666.67$$

$$\sigma = \sqrt{666.67} = 25.82$$

Table 4.2
Computation of Variance and Standard Deviation, Ungrouped Data, Six Attorneys Working on Case A, January 1983

Attorney	Time Involved (x)	$x - \mu$ (μ = mean)	$(x - \mu)^2$	x^2
1	20	−40	1,600	400
2	40	−20	400	1,600
3	60	0	0	3,600
4	60	0	0	3,600
5	80	20	400	6,400
6	100	40	1,600	10,000
Total	360	0	4,000	25,600

Thus, for case A the variance is 666.67, while the standard deviation is 25.82 hours. For case B, using the formulas shown above, the variance is 1,133.33. The standard deviation, on the other hand, is 33.67 hours. One would conclude from the standard deviation in both cases that the dispersion is considerably greater in case B; this conclusion is supported by observation of the data. Stated differently, there is more variation in hours worked on case B as contrasted with case A.

When data are grouped into a frequency distribution, the formulas for the variance and standard deviation are modified. The computational equations for calculating the variance and standard deviation are as shown below:

$$\sigma^2 = \frac{\sum_{i=1}^{K} x_i^2 f_i}{N} - \left(\frac{\sum_{i=1}^{K} x_i f_i}{N}\right)^2 \tag{4.5}$$

$$\sigma = \sqrt{\frac{\sum_{i=1}^{K} x_i^2 f_i}{N} - \left(\frac{\sum_{i=1}^{K} x_i f_i}{N}\right)^2} \tag{4.6}$$

where

x_i = midpoint of the ith class
K = number of classes in the distribution
f_i = frequency of ith class
N = number of observations

To find the variance and standard deviation of the ages of employees in the law firm, first reported in Table 2.3, one would proceed as indicated in Table 4.3, using Equations 4.5 and 4.6:

$$\sigma^2 = \frac{\sum_{i=1}^{K} x_i^2 f_i}{N} - \left(\frac{\sum_{i=1}^{K} x_i f_i}{N}\right)^2$$

$$\sigma^2 = \frac{107{,}100}{50} - \left(\frac{2{,}250}{50}\right)^2 = 117$$

$$\sigma = \sqrt{\sigma^2} = \sqrt{117} = 10.82$$

The standard deviation of the ages of employees in the law firm is 10.82 years. Used in conjunction with the mean or some other measure of central tendency, it is simply a device used to describe the data. In this example, the mean age of employees is 45 with a standard deviation of 10.82 years.

Table 4.3
Computation of Variance and Standard Deviation, Grouped Data, Average Age of Employees, January 1983

Age Group	Midpoint (x)	Frequency (f)	xf	x^2	x^2f
25 but under 35	30	10	300	900	9,000
35 but under 45	40	17	680	1,600	27,200
45 but under 55	50	12	600	2,500	30,000
55 but under 65	60	10	600	3,600	36,000
65 but under 75	70	1	70	4,900	4,900
Total		50	2,250		107,100

RELATIVE DISPERSION

Measures of relative dispersion express the degree of dispersion or variation as a percentage of the average about which they are computed. The primary measure, the coefficient of variation, is normally used to compare the variation in one population with that in another. When it is necessary or desirable to compare two or more different populations with respect to their variation, comparing the standard deviation of the populations may be misleading. This is particularly true if the measurement units of the populations are different or if the magnitudes of the observations vary greatly from distribution to distribution.

For illustrative purposes, assume the following hourly earnings data are presented on two groups of employees working for the same company but in different plants. Based on data on 100 employees in plant 1, the average hourly earnings are $10.45 with a standard deviation of $1.50. In plant 2, average earnings for 125 employees are $12.90 with a standard deviation of $1.80. The labor union representing workers brings suit against plant 2 alleging discrimination based on the larger standard deviation in wages as compared with plant 1.

To measure the relative dispersion between average wages paid in the two plants, the coefficient of variation is an appropriate tool:

$$(4.7) \qquad V = \left(\frac{\sigma}{\mu}\right) = 100$$

where

V = coefficient of variation

Given that $\mu_1 = \$10.45$, $\sigma_1 = \$1.50$, $\mu_2 = \$12.90$, and $\sigma_2 = \$1.80$, the relative dispersion is computed as follows:

$$V_1 = \left(\frac{\sigma_1}{\mu}\right) 100 = \left(\frac{1.50}{10.45}\right) 100 = 14.35\%$$

$$V_2 = \left(\frac{\sigma_2}{\mu}\right) 100 = \left(\frac{1.80}{12.90}\right) 100 = 13.95\%$$

By comparing the two coefficients of variation, one can see that the wage distribution in plant 2 is slightly less variable than that in plant 1.

SKEWNESS

In addition to knowing measures of central tendency and the dispersion in a distribution, it is often helpful to know more about its shape. The third characteristic of a population that helps to complete its description is *skewness*. Skewness describes the lack of symmetry in a distribution. Measures of skewness indicate not only the degree to which a population or data set is skewed but also the direction in which it is skewed. The most common method of measuring skewness makes use of the relative positions of the mean and either the mode or median of a distribution. The degree of skewness of various distributions is illustrated in Figure 4.1.

Skewness results from a preponderance of extreme values at one side or in one tail of a distribution. In a symmetrical distribution, the mean, the median, and the mode are equal. The greater the degree of skewness of a distribution, the greater the difference between these measures, with the arithmetic mean being the furthest in the direction of the extreme values in the longer tail. The mode is unaffected by the extreme values; the median, which is affected by the number but not the values of the extremes, is pulled in the direction of the extremes. The median falls about two-thirds of the distance from the mode to the mean.

The formula for computing the *coefficient of skewness* is presented as Equation 4.8. Since a mode does not always exist, it is more convenient to define this measure using the median:

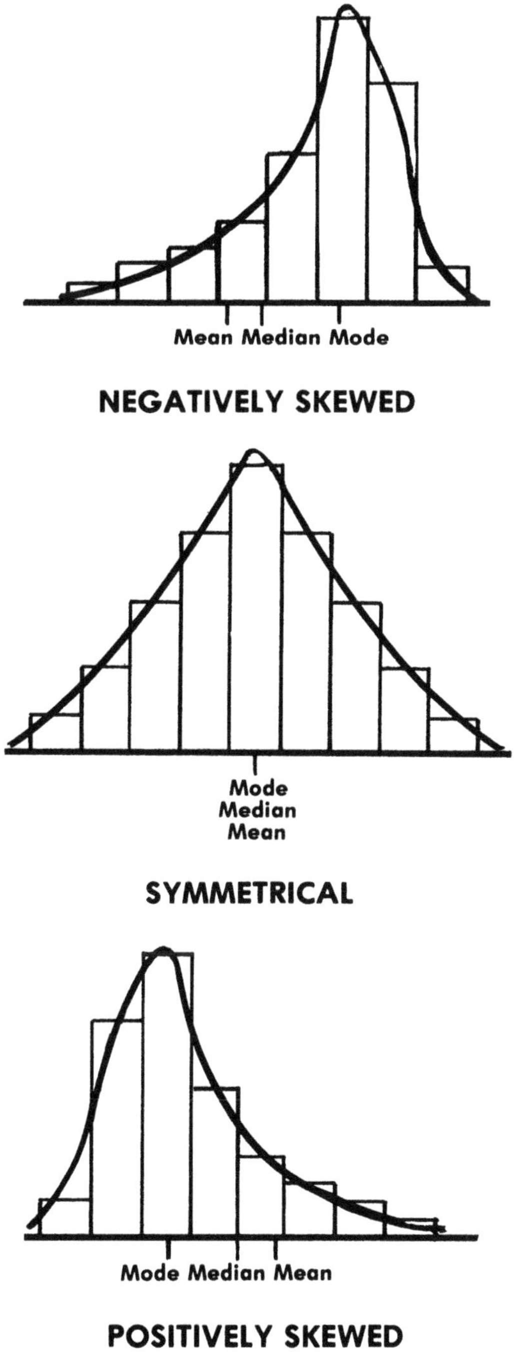

Figure 4.1
Symmetrical and Skewed Distributions

$$(4.8) \qquad sk = \frac{3(\mu - \text{Md})}{\sigma}$$

where

sk = coefficient of skewness
μ = arithmetic mean
Md = median
σ = standard deviation

As noted earlier, when a distribution is symmetrical, the mean and median are equal and $sk = 0$. For positively skewed distributions, the mean is greater than the median, and sk is a positive value. The opposite is true for negatively skewed distributions.

Using the data on ages of employees, the coefficient of skewness can be determined. Remember from this example, $\mu = 45$, Md = 40.6, $\sigma = 10.82$. Thus, using Equation 4.8,

$$sk = \frac{3(\mu - \text{Md})}{\sigma} = \frac{3(45 - 40.6)}{10.82} = 1.22$$

ILLUSTRATIVE USES OF THE STANDARD DEVIATION

At least two landmark decisions handed down by the Supreme Court have made prominent use of the standard deviation. The first of these cases, *Hazlewood School District v. United States,*[1] involved teacher employment discrimination in violation of Title VII of the Civil Rights Act of 1974. The United States brought action against the petitioners—the Hazlewood School District located in St. Louis County, Missouri—alleging that they were engaged in a "pattern or practice" of employment discrimination. The district court ruled that the government had failed to establish a pattern or practice of discrimination.

The court of appeals reversed. In part, the reversal was on the grounds that the trial court's analysis of statistical data rested on an irrelevant comparison of Negro teachers to Negro pupils in Hazlewood instead of a comparison of Negro teachers in Hazlewood to Negro teachers in the relevant labor market area. In the labor market area, 15.4 percent of the teachers were Negro. However, in the 1972–73 and 1973–74 school years, only 1.4 and 1.8 percent, respectively, of Hazlewood's teachers were Negroes. This statistical disparity, particularly when viewed against the background of Hazlewood's teacher hiring procedures, was held to constitute a prima facie case of a pattern or practice of racial discrimination.

The Hazlewood School District, contending that the statistical data on which the court of appeals relied could not sustain a finding of a violation of Title VII, appealed to the Supreme Court. In vacating and remanding the decision of the court of appeals, the Supreme Court held that the former erred in disregarding the statistical data dealing with Hazlewood's hiring after it became subject to Title VII. Further, the appeals courts should have remanded the case to the district court for further findings as to the relevant labor market area and for an ultimate determination as to whether Hazlewood had engaged in a pattern or practice of discrimination since March 24, 1972. Discussing the disparity in hiring of Negroes, the court pointed out the use of the standard deviation:

The percentage of Negroes on Hazlewood's teaching staff in 1972–73 was 1.4%, and in 1973–74 it was 1.8%. By contrast, the percentage of qualified negro teachers in the area was, according to the 1970 census, at least 5.7% . . . the Government contends that a comparative figure of 15.4%, rather than 5.7%, is the appropriate one. . . . But even assuming, *arguendo*, that the 5.7% figure used by the petitioners is correct, the disparity between that figure and the percentage of Negroes on Hazlewood's teaching staff would be more than fourfold for the 1972–1973 school year, and threefold for the 1973–1974 school year. A precise method of measuring the significance of such statistical disparities was explained in *Casteneda v. Partida*. It involves calculation of the "standard deviation" as a measure of predicted fluctuations from the expected value. . . . Using the 5.7% figure as the basis for calculating the expected value, the expected number of Negroes . . . would be roughly 63 in 1972–1973 and 70 in 1973–1974. The observed number in those years was 16 and 22, respectively. The difference between the observed and expected values was more than six standard deviations in 1972–1973 and more than five standard deviations in 1973–1974 . . . "as a general rule for such large samples, if the difference between the expected value and the observed value is greater than two or three standard deviations," then the hypothesis that teachers were hired without regard to race would be suspect.[2]

In *Casteneda v. Partida,*[3] the case cited above, which will be discussed in greater detail in Chapter 6, the court stated that "[t]he measure of the predicted fluctuations from the expected value is the standard deviation . . . in this case the standard deviation is approximately 12."[4] Proceeding, the court stated:

The 11 year data here reflect a difference between the expected and observed number of Mexican-Americans of approximately 29 standard deviations. A detailed calculation reveals that the likelihood that such a substantial departure from the expected value would occur by chance is less than 1 in 10^{140}.[5]

NOTES

1. 433 U.S. 299 (1977).
2. *Id* at 308–09.
3. 430 U.S. 482 (1977).
4. *Id* at 497.
5. *Id.*

SELECTED REFERENCES

Blalock, Hubert M., Jr. *Social Statistics*, Chap. 6. New York: McGraw-Hill Book Co., 1960.

Guenther, William C. *Concepts of Statistical Inference*, Chap. 3. 2d ed. New York: McGraw-Hill Book Co., 1973.

Hoel, Paul G. *Elementary Statistics*, Chap. 2. 2d ed. New York: John Wiley & Sons, 1966.

Miller, Irwin and John F. Freund. *Probability and Statistics for Engineers*, Chap. 6. Englewood Cliffs, N.J.: Prentice-Hall, 1965.

Wallis, W. A. and H. V. Roberts. *The Nature of Statistics*, Chaps. 5, 7. New York: Free Press, 1962.

Wolf, Frank L. *Elements of Probability and Statistics*, Chap. 5. New York: McGraw-Hill Book Co., 1962.

Yamane, Taro. *Statistics: An Introductory Analysis*, Chap. 4. 3d ed. New York: Harper & Row, 1973.

PART II

Coping with Uncertainty

CHAPTER 5

PROBABILITY

This chapter begins the study of statistical inference, a key concept for attorneys. Before beginning the study of statistical inference, however, one must become familiar with the basic concepts of probability. The previous chapters demonstrated that statistical data can be summarized and displayed in various ways. There was no attempt to make any inference beyond the limits of the observed data. To make inferences, one must use probability concepts.

Probability is a most pervasive idea; it creeps into our everyday life at innumerable points. Probability involves the study of uncertain events, and uncertainty characterizes many activities.

THE NATURE OF PROBABILITY

The meaning of probability is not altogether clear. There is disagreement among statisticians, mathematicians, and philosophers about the precise definition of probability. For purposes of this book, probability will be defined as a numerical measure of uncertainty. To be more specific, probability can be defined as a numerical measure of the likelihood of the occurrence of an uncertain event. For example, if an event will never occur, a value of 0 is assigned to the likelihood of this event. If the event will always occur, a value of 1 is assigned. Between the two extremes of "impossibility" and "absolute certainty"—i.e., the numerical values of 0 and 1—lies the probability of the occurrence of a given event. Since there is no likelihood smaller than impossible and nothing is more certain than certainty, a probability measure can never have a value less than 0 or greater than 1.

Objective Probability

Originally, probability was viewed as a purely *objective* measure of relative likelihood. This point of view considers probability as the ratio of successful outcomes of an experiment to the total number of outcomes possible. The experiment—regardless of type—produces facts on which everyone can agree. Using this approach, the probability of any event occurring was considered to be an exact value.

Objective probability, then, is that probability for which there is definite historical evidence and common experience to support the assignment of probabilities. If, for example, out of a group of 1,000 people, 600 have incomes in excess of $15,000, then the probability of a member having an income over $15,000 is .60. Assuming a fair coin, one can, based on historical evidence and common experience, state that the probability of tossing a head is .50. This can be shown to be true by repeatedly tossing a coin. If these conditions are completely known, the probability of an event can be precisely determined. However, in the absence of complete information about these conditions, probabilities must be estimated on the basis of available information.

Subjective Probability

A more recent development is the *subjective* view of probability. According to the subjective view, a probability is simply a number representing a belief concerning the occurrence of a future event, the only limitation being that the number must be between 0 and 1. Stated another way, subjective probability occurs when there is an absence of historical evidence or common experience to support assignment of probabilities.

The objective point of view is used for most of this book. In studying statistical decision theory, the subject of Chapter 8, it is convenient to take the subjective approach. It should be pointed out, however, that once the probability value is determined, regardless of method, the same rules of probability always apply.

RULES OF PROBABILITY

Perhaps the concept of probability can be best illustrated through a Venn diagram. The English logician J. Venn developed this diagram to portray graphically the rules for combining probabilities. To construct a Venn diagram, a space is first enclosed representing the total of all possible outcomes. This space may be any shape and is called a *sample space.*

Consider the following example. A venire (sample space) of 100 people consists of 55 women and 45 men (Figure 5.1). Twenty-two of the women and 15 of the men are college graduates. In formal statistical terms, the group is called a sample space which includes 100 elements. The sample space or universal set, *U*, can be broken down into four subsets—men, women, college graduates, and nongraduates. The subsets men and women are nonintersecting and may be defined as being *mutually exclusive;* the occurrence of one precludes the occurrence of the other. In practical terms, this means that there can be no single element that has both the characteristic male and female. The same is true of the subsets college graduate and nongraduate. Other subsets, such as male and college graduate, are not mutually exclusive since one person can possess both

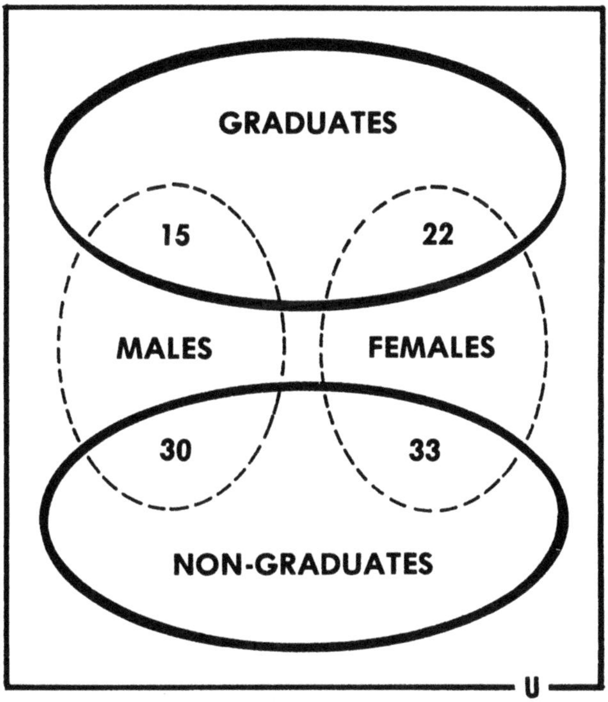

Figure 5.1
Venn Diagram, Education Level and Sex

Table 5.1
Frequency Table, by Level of Education and Sex, 100 Veniremen, January 1983

Education Level	Male (*M*)	Female (*F*)	Total
College Graduate (*C*)	15	22	37
Nongraduate (*N*)	30	33	63
Total	45	55	100

characteristics. The situation described above can be summarized in a frequency table such as Table 5.1.

Marginal Probability

Suppose an attorney wishes to know the probability of selecting a male juror, assuming every person has an equal chance of being chosen. Since 45 of the 100 people in the venire are male, the probability of a male venireman being chosen at random is .45. In general terms, this statement may be summarized as follows:

$$(5.1) \qquad P(A) = \frac{n(A)}{n(U)}$$

where

$P(A)$ = probability of event A occurring
$n(A)$ = number of elements in subset A
$n(U)$ = number of elements in universal set U

Thus,

$$P(A) = \frac{45}{100} = .45$$

This probability is called *marginal probability*. The term *marginal* relates to the fact that the number in the numerator is found at the margin of the frequency table. Thus, the probability of any event that includes the number of elements in a single category of classification—a single subset—as the numerator and the number of elements in the universal set in the denominator is called marginal probability. For example, the probability of selecting a college graduate, C, is

$$P(C) = \frac{n(C)}{n(U)} = \frac{37}{100} = .37$$

Joint Probability

Joint probability refers to the occurrence of two or more events together. What is the probability of selecting a juror who is male and a college graduate? This can be determined quite easily:

$$(5.2) \qquad P(A \text{ and } B) = \frac{n(A \text{ and } B)}{n(U)}$$

where

$P(A \text{ and } B)$ = probability of events A and B occurring together
$n(A \text{ and } B)$ = number of elements in subsets A and B

The probability of randomly selecting a juror who is male, *m*, and a college graduate is .15:

$$P(M \text{ and } C) = \frac{n(M \text{ and } C)}{n(U)} = \frac{15}{100} = .15$$

In a like manner, the probability of selecting a juror who is female and a college graduate is .22.

Conditional Probability

Problems frequently arise that deal with the probabilities for only a portion of the sample space. The probability of an event depends upon what portion of the sample space is considered. Suppose a male is selected from the venire. What is the probability that he is a college graduate? Looking at the frequency table, one determines that 15 of 45 males in the sample space are college graduates. Therefore, the probability that the person selected would be a college graduate *given* that he was a male is .33. This is determined as follows:

(5.3) $$P(A|B) = \frac{n(A \text{ and } B)}{n(B)}$$

where

$P(A|B)$ = probability of event A occurring given that event B has occurred
$n(A \text{ and } B)$ = number of A and B events
$n(B)$ = number of *all* B events

$$P(C|M) = \frac{n(C \text{ and } M)}{n(M)} = \frac{15}{45} = .33$$

Similarly, given a female has been selected, the probability that she will not be a college graduate is

$$P(N|F) = \frac{n(N \text{ and } F)}{n(F)} = \frac{33}{55} = .60$$

The General Law of Multiplication

The *general law of multiplication* applies when probabilities are conditional. In fact, the law of multiplication is simply a restatement of the definition of conditional probability.

(5.4) $$P(A \text{ and } B) = P(A) \times P(B|A)$$

Table 5.2 is a restatement, in terms of probabilities, of the Venn diagram and Table 5.1. Using this table, suppose the attorney wanted to know the probability

Table 5.2
Probability Table, by Level of Education and Sex, 100 Veniremen, January 1983

Education Level	Male (M)	Female (F)	Total
College Graduate (C)	.15	.22	.37
Nongraduate (N)	.30	.33	.63
Total	.45	.55	1.00

of selecting two successive jurors at random who are males and college graduates? Using Equation 5.4:

$$P(MC_1 \text{ and } MC_2) = P(MC_1) \times P(MC_2|MC_1)$$

$$P(MC_1 \text{ and } MC_2) = 15/100 \times 14/99$$

$$= .021$$

The marginal probability of selecting a male who is a college graduate is .15 since 15 of 100 people in the venire are males with college degrees. The conditional probability of selecting a male college graduate on the second draw *given* a male college graduate on the first draw is 14/99 since there are 14 college graduates left among the remaining 99 people. In a similar manner, the probability of selecting two females in succession who are not college graduates is:

$$P(FN_1 \text{ and } FN_2) = P(FN_1) \times P(FN_2|FN_1)$$

$$P(FN_1 \text{ and } FN_2) = (33/100)\ (32/99) \doteq .107$$

The general rule of multiplication is applicable when one desires to find the probability that two events (say, events A and B) will both occur given that one event (say, event A) has already occurred. The rule also applies to more than two draws. For example, the probability of selecting in sequence three male college graduates is .00279.

The Special Law of Multiplication: Independent Events

If the outcome of one event in no way affects the outcome of another event, the outcomes are considered to be *independent*. This statistical independence can be illustrated by the toss of two coins. The outcome of one of the coins (head or tail) is unaffected by the outcome of the other coin. Thus, the probability of a tail occurring on two successive tosses of a coin is the product of the marginal probability of a tail occurring on each toss, i.e., $P(T_1 \text{ and } T_2) = P(T_1) \times P(T_2) = .5 \times .5 = .25$.

An event B is said to be independent of an event A if $P(B|A) = P(B)$. In the case of statistical independence, the law of multiplication reduces to

(5.5) $P(A \text{ and } B) = P(A) \times P(B)$

From Table 5.2, it is obvious that educational level is not independent of sex. If it were, the product of the marginals would equal the corresponding joint probabilities. In this table, the product of male and college graduate, $.45 \times .37$, does not equal the joint probability of .15.

The General Law of Addition

The *law of addition* is used to calculate the probability that a number of events will occur either separately or jointly. To illustrate, one might ask what the probability is that a card selected at random from a standard deck of cards will be either a jack or a diamond. There is, for instance, some chance that the card chosen will be both a jack and a diamond, e.g., the jack of diamonds. That is, there is an overlapping of outcomes. This means that both A (a jack) and B (a diamond) can occur together. The joint probability of this occurring is $P(A \text{ and } B)$ as noted earlier. It seems logical that the joint probability of a jack and a diamond should be subtracted to avoid double counting. Otherwise, it will be added twice in finding the probability that a card chosen from a deck will be a jack or a diamond. The general law of addition is:

(5.6) $P(A \text{ or } B) = P(A) + P(B) - P(A \text{ and } B)$

In this example, $P(A) = 4/52$, $P(B) = 13/52$, and $P(A \text{ and } B) = 1/52$. Thus, the probability of choosing a jack or a diamond is 16/52 or .3077.

To relate the general law of addition to the jury selection process, an attorney may want to know the probability of selecting a college graduate or a male from the venire. Using Equation 5.6 and Table 5.2, the probability becomes

$$P(M \text{ or } C) = P(M) + P(C) - P(M \text{ and } C) = (.45) + (.37) - (.15) = .67$$

The Special Law of Addition

If two events are mutually exclusive, there are no elements common to both events. In such a case, $P(A \text{ and } B) = 0$, and the general rule of addition reduces to the special case:

(5.7) $P(A \text{ or } B) = P(A) + P(B)$

This rule can be illustrated in the venire example. Since males (M) and females (F) are mutually exclusive,

$$P(M \text{ or } F) = P(M) + P(F) = .45 + .55$$

$$P(M \text{ or } F) = 1.0$$

On the other hand, the probability of selecting a male and a female equals 0.

This can be further illustrated by rolling a die. The possible outcomes of one roll, all mutually exclusive, are listed below:

Outcome	*Probability*
1	1/6
2	1/6
3	1/6
4	1/6
5	1/6
6	1/6
Total	1

Thus, using the special rule of addition, the probability of rolling a 1 or a 6 is 1/3 as shown below:

$$P(1 \text{ or } 6) = P(1) + P(6) = 1/6 + 1/6 = 1/3$$

Bayes' Theorem

In the eighteenth century, a Scottish pastor, the Reverend Thomas Bayes, manipulated the laws of addition and multiplication to form a rule which has proved to be the cornerstone of modern decision theory (the subject of Chapter 8). Bayes' theorem, as this rule is most often called, makes it possible to determine a conditional probability in terms of the complementary or reverse conditional probability and the relevant marginal probabilities.

This involves the use of *prior* probabilities, determined before any experimental information is available, and *posterior* probabilities, which are determined after the result of an experiment is known. The basic principle of Bayesian analysis is that additional information (if available) can sometimes enable individuals to alter (improve) the marginal probabilities of the occurrence of an event. The altered probabilities are referred to as revised or posterior probabilities.

Bayes' theorem includes two kinds of information. $P(A_i)$ represents the *prior* probabilities associated with certain events. These represent the *prior* information. The *sample* or *experimental* information is represented by B, an event referred to as the statistical sample, and $P(B|A_i)$, the probability of B given the occurrence of some event A_i.

From the definition of conditional probability (Equation 5.3),

$$P(A_i|B) = \frac{P(A_i \text{ and } B)}{P(B)}$$

$P(A_i$ and $B)$ is a joint probability (Equation 5.4) which may be rewritten as

$$P(A_i \text{ and } B) = P(A_i) \times P(B|A_i)$$

This is the multiplication rule for two events that are dependent.

$P(B)$ is a marginal probability, which can represent the sum of a series of joint probabilities. Therefore, $P(B)$ can be rewritten in expanded form as

$$(5.8) \qquad P(B) = P(A_1 \text{ and } B) + P(A_2 \text{ and } B) + P(A_i \text{ and } B).$$

Substituting Equation 5.4 into Equation 5.8 yields Equation 5.9:

$$(5.9) \qquad P(B) = (PA_1) \times P(B|A_i) + P(A_2) \times P(B|A_2) + \ldots + P(A_i) \times P(B|A_i)$$

or, in more compact form,

$$(5.10) \qquad P(B) = \sum_{i=1}^{n} P(A_i) \times P(B|A_i)$$

Thus, Equation 5.10 represents the marginal probability of event B.

The final step in developing Bayes' theorem is to substitute Equations 5.4 and 5.10 into Equation 5.3. Thus,

$$(5.11) \qquad P(A_i|B) = \frac{P(A_i) \times P(B|A_i)}{\sum_{i=1}^{n} P(A_i) \times P(B|A_i)}$$

Note that both the prior information, $P(A_i)$, and the sample information, $P(B|A_i)$, are included in the equation.

The following industrial example will be used initially to illustrate Bayes' theorem. A production manager for a manufacturing firm is supervising the machine setup for a product to be produced. A machine operator sets up the machine. When the machine is set up correctly by an operator, there is a 10 percent chance that an item produced on the machine will be defective. If the machine is set up incorrectly, there is a 70 percent chance the item will be defective. Based on past experience, the production manager knows there is a 25 percent chance that the machine will be set up incorrectly by the operator. To reduce the chance that an item produced on a machine will be defective, the manager decides to have the operator produce a sample item. If the sample item turns out to be defective, what is the probability the machine is set up incorrectly?

The above probabilities can be summarized in Bayesian form as follows:

$$P(A_1) = .25 \qquad P(B|A_1) = .70$$

$$P(A_2) = .75 \qquad P(B|A_2) = .10$$

where

A_1 = correct setup
A_2 = incorrect setup
B = defective part

The posterior probability is the conditional probability that the machine has been set up correctly, given the sample item proves to be defective, or $P(A_1|B)$. Hence, using Equation 5.11:

$$P(A_1|B) = \frac{(.25)(.70)}{(.25)(.70) + (.75)(.10)} = \frac{.175}{.175 + .075} = \frac{.175}{.250} = .70$$

Previously, the production manager knew only that there was a 25 percent chance that the machine had been set up incorrectly. By producing a sample item and testing, the manager now knows there is a .70 probability the machine was set up incorrectly. Thus, by collecting additional information, the manager is able to revise the probability estimate concerning the correct setup of the machine.

As a simple legal illustration of Bayes' theorem (which will be addressed at length in Chapter 8), assume an attorney is questioning prospective jurors. Based on his subjective estimates, the attorney believes that there is a 40 percent chance that any person selected at random possesses the characteristics he desires in a juror for this particular case. In the selection process, the attorney has developed one question which he has been told has proved moderately successful as a predictor of the traits he desires in this situation. With the assistance of a statistician he has retained in the case, the attorney knows there is an 80 percent chance the prospective juror has the requisite characteristics if he answers the question affirmatively. However, there is a 15 percent chance he has the requisite characteristics if he responds in the negative. Assuming the person answers in the affirmative, what is the revised probability he would be acceptable based on the characteristics the attorney seeks?

The relevant probabilities are:

$P(A_1) = .40$ $\quad\quad$ $P(B|A_1) = .80$

$P(A_2) = .60$ $\quad\quad$ $P(B|A_2) = .15$

Using Equation 5.11, the revised probability is computed as follows:

$$P(A_1|B) = \frac{(.40)(.80)}{(.40)(.80) + (.60)(.15)} = \frac{.32}{.32 + .09}$$

$$P(A_1|B) = .78$$

Based on the posterior probability estimate, it would appear that this person has a 78 percent chance of possessing the required characteristics. This is higher than the 40 percent figure assumed for a person selected at random.

ILLUSTRATIVE USES OF PROBABILITY THEORY

A case which illustrates statistical awareness of probability theory on the part of the courts—even though evidence based on mathematical probability was said to be insufficient—is *People v. Collins.*[1] In *People v. Collins,* an elderly woman walking home in an alley in the San Pedro area of Los Angeles was assaulted from behind and robbed. The victim said that she managed to see a young woman with blond hair run from the scene. Another witness, a man who lived on the street at the end of the alley, said that a Caucasian woman with dark blond hair and a ponytail ran out of the alley and entered a yellow automobile parked across the street from him. According to this witness, the automobile was driven by a black male with a mustache and a beard. A few days later, officers investigating the robbery arrested a couple on the strength of these descriptions and charged them with the offense.

At their trial, the prosecution called as an expert witness an instructor of mathematics at a state college in an attempt to establish that the probability was overwhelming that the accused were guilty because they answered this unusual description. His computations assumed that the robbery was committed by a Caucasian with a ponytail who left the scene in a yellow automobile driven by a black man with a beard and a mustache. The witness testified to the product rule of elementary probability theory—what has been in this book described as the special law of multiplication (Equation 5.5). After the prosecutor had him assume individual probabilities for the relevant characteristics, the witness, applying the product rule, concluded that there would be but 1 chance in 12 million that a couple selected at random would possess the incriminating characteristics. The jury convicted the defendants. On appeal, the Supreme Court of California reversed the conviction, holding that the trial court should not have admitted the evidence pertaining to probability:

> We deal here with the novel question whether evidence of mathematical probability has been properly introduced and used by the prosecution in a criminal case. While we discern no inherent incompatibility between the disciplines of law and mathematics and intend no general disapproval or disparagement of the latter as an auxiliary in the fact-finding processes of the former, we cannot uphold the technique employed in the instant case . . . the testimony as to mathematical probability infected the case with fatal error. . . . Mathematics, a veritable sorcerer in our computerized society, while assisting the trier of fact in the search for truth, must not cast a spell over him.[2]

That probability theory was classically misused in this case is readily apparent to one trained in the discipline. As the court stated,

[w]*ithout presenting any statistical evidence whatsoever in support of the probabilities for the factors selected,* the prosecutor then proceeded to have the witness *assume* probability factors for the various characteristics which he deemed to be shared by the guilty couple and all other couples answering to such distinctive characteristics.[3]

Applying the special rule of multiplication to the six factors listed below and the probability of the occurrence of each event—all suggested by the prosecutor—the mathematician stated the probability was 1 in 12 million that any couple possessed the distinctive characteristics of the defendants.

Characteristic	*Individual Probability*
Yellow automobile	$^1/_{10}$
Man with mustache	$^1/_4$
Girl with ponytail	$^1/_{10}$
Girl with blond hair	$^1/_3$
Negro man with beard	$^1/_{10}$
Interracial couple in car	$^1/_{1.000}$

Under this theory, it was to be inferred that there would be but 1 chance in 12 million that the defendants were innocent and that another equally distinctive couple actually committed the robbery.

The court noted that the mathematician's testimony suffered from two basic and pervasive defects—an inadequate evidentiary foundation and an inadequate proof of statistical independence. Addressing the first defect, the court found "the record devoid of any evidence relating to any of the six individual probability factors used by the prosecutor and ascribed by him to the six characteristics."[4] To put it another way, the court stated that the prosecution produced no evidence whatsoever showing, or from which it could be in any way inferred, that only one out of every ten cars which might have been at the scene of the robbery was partly yellow. Nor was there any evidence to show that only one out of every four men who might have been there wore a mustache.

Turning to the issue of statistical independence of the six factors, the court noted that "[n]o proof was presented that the characteristics selected were mutually independent, even though the witness himself acknowledged that such condition was essential to the proper application of the 'product rule' or 'multiplication rule.' "[5] To the extent that the traits were not independent (e.g., blacks with beards and men with mustaches obviously represent overlapping categories), the multiplication rule would inevitably yield a wholly erroneous and exaggerated result even if all of the individual components had been determined with precision.

People v. Collins is not representative of statistical presentations; in fact, its pseudostatistics cannot be taken seriously. The method used in the case, however, was representative of more sophisticated efforts made in earlier cases in which experts sought to make the identification on the basis of similarities in typewriting, handwriting, fibers, or hairs.[6]

NOTES

1. 438 P. 2d 33 (1968).
2. *Id.*
3. *Id* at 36-7.
4. *Id* at 38.
5. *Id* at 39.
6. Miller v. State, 399 S. W. 2d 268 (1966); People v. Jordan, 45 Cal. 2d 697, 290 P. 484 (1955); People v. Trujillo, 32 Cal. 2d 105, 194 P. 2d 681, *Cert. denied,* 355 U.S. 887 (1948); State v. Sneed, 76 N. M. 349, 414 P. 2d 858 (1966).

SELECTED REFERENCES

Ayer, A. J. "Chance." *Scientific American,* Vol. 213, No. 4 (October 1965), pp. 44–54.

Brilmayer, Lea and Lewis Kornhauser. "Review: Quantitative Methods and Legal Decisions." *University of Chicago Law Review,* Vol. 46 (1978), pp. 116–53.

Campbell, Stephen K. *Flaws and Fallacies in Statistical Thinking,* Chap. 11. Englewood Cliffs, N.J.: Prentice-Hall, 1974.

Cramer, Harold. *The Elements of Probability Theory.* New York: John Wiley & Sons, 1955.

Drake, Alvin W. *Fundamentals of Applied Probability Theory.* New York: McGraw-Hill Book Co., 1967.

Finkelstein, Michael O. and William B. Fairley. "A Bayesian Approach to Identification Evidence." *Harvard Law Review,* Vol. 83 (1970), pp. 489–517.

———. "A Comment on 'Trial by Mathematics.'" *Harvard Law Review,* Vol. 84 (1971), pp. 1801–25.

Freund, John E. *Mathematical Statistics,* Chaps. 1–8. 2d ed. Englewood Cliffs, N.J.: Prentice-Hall, 1971.

Kaye, David. "Probability Theory Meets Res Ipsa Loquitur." *Michigan Law Review,* Vol. 77 (June 1979), pp. 1456–84.

Lempert, Richard O. "Modeling Relevance." *Michigan Law Review,* Vol. 75 (April–May 1977), pp. 1021–57.

Miller, Irwin and John E. Freund. *Probability and Statistics for Engineers,* Chap. 2. Englewood Cliffs, N.J.: Prentice-Hall, 1965.

Tribe, Lawrence H. "Trial by Mathematics: Precision and Ritual in the Legal Process." *Harvard Law Review,* Vol. 84 (1971), pp. 1329–74.

Wolf, Frank L. *Elements of Probability and Statistics.* New York: McGraw-Hill Book Co., 1962.

CHAPTER 6

THE BINOMIAL DISTRIBUTION

Determining probabilities for various types of events can be simplified by establishing a general model for the sample space of the problem. A model used to obtain probabilities is called a probability distribution. A model is a simplified representation of something. Mathematical models use mathematical symbols to represent real objects or processes; probability functions are mathematical models. The accuracy of the probability function in estimating probabilities depends on how well the model fits the situation.

There are several well-known probability distributions that are suitable for a wide range of applications. This chapter is concerned with the *binomial distribution*. Before examining this distribution, however, the term *random variable* must be considered.

RANDOM VARIABLE

All probability distributions deal with random variables, those whose value depends entirely on chance rather than choice. The value of a random variable is the general outcome of a random (or probability) experiment. It is useful to distinguish between the random variable itself and the values that it can take on. It is important to note that the value of a random variable is unknown until the event occurs, i.e., until the random experiment has been performed. However, the probability that the random variable will be any specific value is known in advance. For example, if x is defined as a random variable for the experiment of rolling one die, then the value of x may be 1, 2, 3, 4, 5, or 6 on any roll selected at random since each of these outcomes is equally likely. A *frequency function* is the frequency with which each value of x is observed to occur. Thus, for this experiment the frequency of x, f(x), would be

Sample Space = [1, 2, 3, 4, 5, 6]

Frequency, f(x): [$\frac{1}{6}$, $\frac{1}{6}$, $\frac{1}{6}$, $\frac{1}{6}$, $\frac{1}{6}$, $\frac{1}{6}$]

The distribution of the frequencies is the distribution of the probabilities for each of these occurrences.

In the above experiment, the possible values of x are limited to certain numbers. When the random variable is limited to certain values, it is called a *discrete random* variable. This type of variable can assume only certain values, often integers, and no intermediate value can be taken on. For example, the number of persons arriving at a service counter per hour or the number of automobiles sold during a month is a discrete random variable that can be represented by counted data. An arrangement of the probabilities associated with the values of a discrete random variable is called a *discrete probability distribution.*

There are experiments, however, in which the random variable is not limited to certain discrete values but can assume any value within a range of values. If the possible values of x constitute a continuous sample space, then the frequency function associated with these values is continuous. The values of a *continuous random variable* are obtained by measuring from a continuous scale. Measures such as weight, time, distance, and volume involve measurement on a continuous scale. This type of scale has an infinite number of possible values that need not be whole numbers but that may be carried out to as many decimal places as the accuracy of measurement will justify. An arrangement of the probabilities associated with the groups of values in intervals of a continuous random variable is called a *continuous probability distribution*—the subject of Chapter 7.

Expected Value of a Random Variable

The expected value of a random variable is the sum of the values of the random variable weighted by the probability that the random variable will take on that value. Therefore, the expected value is the mean or weighted average. Consider, for example, the possible number of heads that could occur on five tosses of a fair coin (Table 6.1 and Figure 6.1). Probabilities of the values of the random variable, x, range from .03125 (for 0 or 5 heads) to .3125 (for 2 or 3 heads). Of course, the sum of the probabilities must equal 1.0. The expected value of a random variable is calculated by weighting each value of the random variable by its probability and summing:

$$E(x) = \sum_{i=1}^{n} x_i P(x_i) \quad (6.1)$$

where

x_i = ith value of the random variable
$P(x_i)$ = probability of the ith value

In the fair coin example, $E(x)$ is calculated as follows:

$$E(x) = \sum_{i=1}^{5} (0)(.03125) + (1)(.15625) + (2)(.3125) + (3)(.3125)$$

$$+ (4)(.15626) + (5)(.03125)$$

$$E(x) = 2.5$$

able 6.1
robabilities for the Possible Outcomes of an Experiment
onsisting of Five Tosses of a Fair Coin

Values of andom Variable (x)	Outcomes	Probability of x ($P(x)$)
0	(TTTTT)	$(\frac{1}{2} \cdot \frac{1}{2} \cdot \frac{1}{2} \cdot \frac{1}{2} \cdot \frac{1}{2}) = .03125$
1	(HTTTT) (THTTT) (TTHTT) (TTTHT) (TTTTH)	$5(\frac{1}{2} \cdot \frac{1}{2} \cdot \frac{1}{2} \cdot \frac{1}{2} \cdot \frac{1}{2}) = .15625$
2	(HHTTT) (HTHTT) (HTTHT) (HTTTH) (THHTT) (THTHT) (THTTH) (TTHHT) (TTHTH) (TTTHH)	$10(\frac{1}{2} \cdot \frac{1}{2} \cdot \frac{1}{2} \cdot \frac{1}{2} \cdot \frac{1}{2}) = .31250$
3	(HHHTT) (HHTHT) (HHTTH) (TTHHH) (THTHH) (THHHT) (THHTH) (HTHHT) (HTHTH) (HTTHH)	$10(\frac{1}{2} \cdot \frac{1}{2} \cdot \frac{1}{2} \cdot \frac{1}{2} \cdot \frac{1}{2}) = .31250$
4	(HTHHH) (HHTHH) (HHHTH) (HHHHT) (THHHH)	$5(\frac{1}{2} \cdot \frac{1}{2} \cdot \frac{1}{2} \cdot \frac{1}{2} \cdot \frac{1}{2}) = .15625$
5	(HHHHH)	$(\frac{1}{2} \cdot \frac{1}{2} \cdot \frac{1}{2} \cdot \frac{1}{2} \cdot \frac{1}{2}) = .03125$
	Total	1.00000

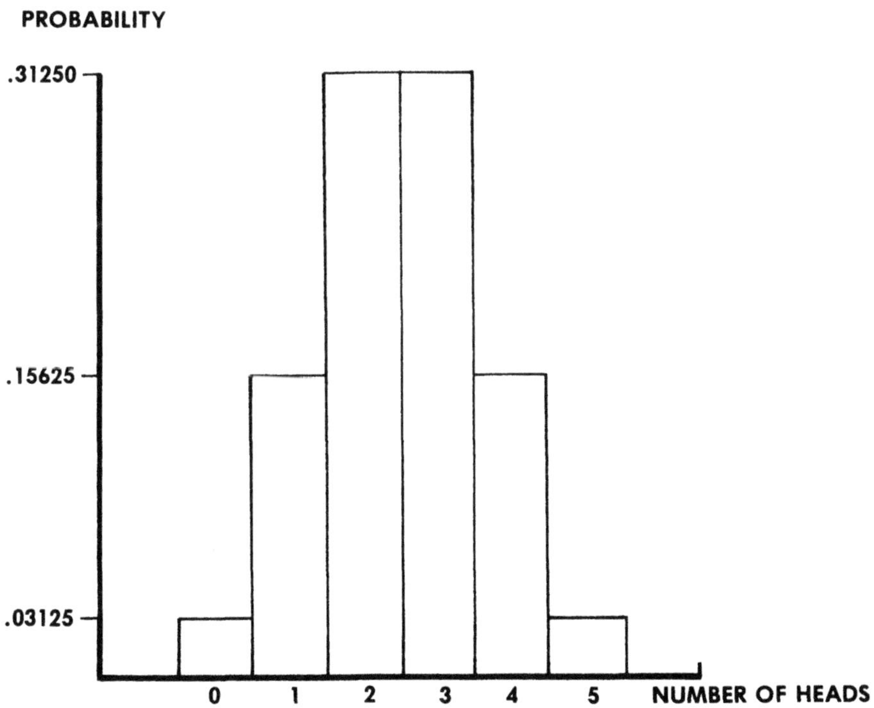

Figure 6.1
Discrete Probability Distribution of the Experiment Involving Five Tosses of a Fair Coin

Table 6.2
Computation of the Expected Value, Number of Heads, Five Tosses of a Fair Coin

Values of Random Variable (x)	Probability of x (P(x))	Expected Value of x (E(x))
0	.03125	.10
1	.15625	.15625
2	.31250	.62500
3	.31250	.93750
4	.15625	.62500
5	.03125	.15625
	1.00000	$E(x)$ = 2.50000

Thus, the expected or average number of heads on five tosses of a fair coin is 2.5. This is also shown in Table 6.2.

Variance and Standard Deviation of a Random Variable

Measures of variation or dispersion were discussed in Chapter 4. The most useful of these measures are the *variance* and the *standard deviation*. The same measures are useful to help describe probability distributions of random variables. The variance of the distribution of a random variable, the square of the standard deviation, is defined as:

$$\text{(6.2)} \qquad \text{Var}\ (x) = \sum_{i=1}^{n} (x_i - E(x))P(x_i)$$

where

x_i = *i*th value of the random variable
$E(x)$ = mean
$P(x_i)$ = probability of *i*th random variable

The variance of the random variable associated with number of heads occurring on five tosses of a fair coin is 1.76 (Table 6.3).

BINOMIAL PROBABILITY

One of the most common discrete probability distributions is the *binomial distribution*. This distribution results from repeated independent trials of an experiment in which there are only two possible mutually exclusive outcomes on each trial. The binomial distribution has the following characteristics:

1. The experiment can be viewed as a sequence of separate and distinct trials.
2. Each trial could result in either of two outcomes—success or failure.
3. The probability of success on any given trial remains constant throughout the sequence of trials.
4. The trials are statistically independent, i.e., the outcome of any given trial does not depend on the outcome of other trials in the sequence.

Trials which meet the conditions set forth in conditions 2 through 4 are called *Bernoulli trials*. Binomial probabilities measure the probability of obtaining a given number of successes in a given number of trials. Successive tosses of a particular coin represent Bernoulli trials; the probability of obtaining eight heads out of ten tosses is a binomial probability. The *binomial probability distribution* is a systematic arrangement of the probabilities associated with the possible number of successes associated with a given number of trials.

Examples of Bernoulli processes and uses of binomial probabilities are by no means limited to coin-tossing or dice-rolling experiments. There are many empirical processes of importance which are classified as being binomial in nature. For example, each unit coming off an assembly line might be classified as defective or not defective. Vouchers examined by an auditor may be viewed as correct or incorrect.

It should be noted that many processes can be characterized as Bernoulli for analytical purposes even though they are not true Bernoulli processes in every respect. If the ''fit'' is close enough, it may be assumed that the Bernoulli process is a reasonable characterization. For example, a survey similar to that noted in *Bristol Myers v. Federal Trade Commission* might be used to determine whether consumers prefer liquid to bar soaps. The outcome of a survey interview could be characterized as ''yes'' (success) or ''no'' (failure) answers to the question. If the sample of consumers was random, Bernoulli might be a useful description of the process.

Table 6.3
Computation of the Variance, Number of Heads, Five Tosses of a Fair Coin

Values of Random Variable (x)	Probability of x ($P(x)$)	$(x - E(x))^2$	$(x - E(x))^2P(x)$
0	.03125	6.25	.1953125
1	.15625	2.25	.3515625
2	.31250	.25	.0781250
3	.31250	.25	.0781250
4	.15625	2.25	.3515625
5	.03125	6.25	.1953125
	$E(x) = 2.5$		$Var(x) = 1.7578125$

Very often the effect of departures from theoretical conditions on the solutions of practical problems is minor. Furthermore, the assumption of Bernoulli processes may make difficult problems tractable. In any case, uses of binomial probabilities are so widespread as to justify particular attention to their calculation.

Calculation of Binomial Probabilities

Application of the binomial distribution to administrative problems often involves an analysis of data obtained from sampling. In drawing samples of outcomes from a Bernoulli process, the results—expressed as numbers of successes and failures—can vary from sample to sample. This is illustrated by the example in Table 6.1. In this example, there are 32 possible sequences of successes and failures (H and T). In calculating binomial probabilities, it is important to determine how many possible sequences correspond to the given value of a discrete random variable.

The probability of r successes in a particular order in n trials is

$$(p)^r\,(1 - p)^{n - r}$$

where

$$\begin{aligned} n &= \text{number of trials} \\ r &= \text{number of successes in } n \text{ trials} \\ n - r &= \text{number of failures in } n \text{ trials} \\ p &= \text{probability of success in any trial} \\ (1 - p) &= \text{probability of failure in any trial} \end{aligned}$$

The number of possible orderings of successes and failures is given by the combinatorial formula for the combination of n elements taken r at a time:

$$C(n,r) = \frac{n!}{r!(n - r)!} \qquad (6.3)$$

The symbol ! stands for *factorial*. For $n = 3$, $n!$ means $3 \times 2 \times 1$. $0!$ is always 1.

The application of combinations to binomial probability distributions is illustrated by the example shown in Table 6.1 and Figure 6.1. The combinations are the number of possible outcomes. For example, the combination for which there are no heads in five tosses of a fair coin is the sequence (TTTTT). As shown in Table 6.1, there are five ways to obtain one head in five tosses of the coin. Using Equation 6.3 this is calculated as follows:

$$C(5,1) = \frac{5!}{1!(5 - 1)!} = \frac{5!}{(1)(4!)} = \frac{5 \times 4 \times 3 \times 2 \times 1}{1 \times 4 \times 3 \times 2 \times 1} = \frac{120}{24} = 5$$

Similarly, there are ten ways to obtain two heads in five tosses:

$$C(5,2) = \frac{5!}{2!(5-2)!} = \frac{5!}{(2!)(3!)} = \frac{5 \times 4 \times 3 \times 2 \times 1}{(2 \times 1)(3 \times 2 \times 1)} = \frac{120}{12} = 10$$

The generalized formula for binomial probability can now be written by combining the term describing the probability of the desired outcome in a particular order with the combinatorial term which counts the number of possible orderings:

$$(6.4) \qquad P(r|n,p) = \frac{n!}{r!(n-r)!} p^r(1-p)^{n-r}$$

This is read as the binomial probability of r successes in n trials given the probability of success p on any given trial.

To demonstrate the use of this formula, consider the following examples. The probability of obtaining three sixes in four tosses of a die is:

$$P(r = 3|n = 4,\ p = 1/6) = \frac{4!}{3!(4-3)!}(1/6)^3(5/6)^1$$

$$= \frac{4!}{(3!)(!)}(1/216)(5/6)$$

$$= 4(5/1296) = .0154$$

On the other hand, the probability of obtaining exactly eight heads out of ten tosses of a fair coin is:

$$P(r = 8|n = 10,\ p = .50) = \frac{10!}{8!(10-8)!}(\tfrac{1}{2})^8(\tfrac{1}{2})^2$$

$$= \frac{10!}{8!(2!)}(1/256)(\tfrac{1}{4})$$

$$= .044$$

THE BINOMIAL DISTRIBUTION

The binomial distribution is a family of probability distributions describing the probabilities of the possible outcomes for all possible combinations of n and p. Each of these distributions is defined by a unique combination of n and p. Since n can be any positive integer and p any value between 0 and 1, the size of the family is unlimited. The values n and p which determine a particular binomial distribution are called the *parameters* of the distribution. To put this in the proper perspective, the probability equality and inequality symbols and relationships presented in Table 6.4 are useful. For illustrative purposes, assume r is 8, n is 10, and p is .50.

To illustrate, consider the binomial probabilities (Equation 6.4) for a particular

Table 6.4
Binomial Probability Equality and Inequality Symbols and Relationships

Relationship or Symbol	Interpretation
$P(r = 8 \mid n = 10, p = .50)$	Probability of r successes, given a specific number n of trials and some specific probability (p) of success.
$P(r \geq 8 \mid n = 10, p = .50)$	Probability that the number of successes is *greater than or equal to* a specific number, r, given values for n and p.
$P(r > 8 \mid n = 10, p = .50)$	Probability that the number of successes is *greater than* a specific number, e.g., 8.
$P(r \leq 8 \mid n = 10, p = .50)$	Probability that the number of successes is *less than or equal to* a specific number.
$P(r < 8 \mid n = 10, p = .50)$	Probability that the number of successes is *less than* a specific number.

member of this family with parameters $n = 5$ and $p = .50$, the combination describing the five tosses of the fair coin discussed earlier. For example,

$$P(r = 0 \mid n = 5, p = .50) = \frac{5!}{0!(5-0)!}(.50)^0(.50)^5$$

$$= \frac{5!}{(0!)(5!)}(1)(.03125)$$

$$= .03125$$

Probabilities for the other values of r are calculated in a like manner and are

Table 6.5
Binomial Probability Distribution for $n = 5$, $p = .50$

Number of Successes (r)	Probability of r
0	.03125
1	.15625
2	.31250
3	.31250
4	.15625
5	.03125
Total	1.00000

summarized in Table 6.5. This summarization of these results for all possible outcomes $r = 0$ through $r = 5$ is called a binomial probability distribution.

Binomial Tables

As can be seen from the above, calculation of binomial probabilities is a tedious and time-consuming task. Fortunately, many binomial probabilities have been calculated for various values of p and n and published in tables. Although the computation of binomial probability is conceptually straightforward, the arithmetic involved in many practical problems could be virtually insurmountable. Thus, binomial tables save considerable time and effort.

Cumulative probabilities compound the calculation. For example, the probability of *at least* 2 prospective jurors out of 20 having birthdays on a particular day is a cumulative probability. At least 2 people means 2 or more. To calculate the probability of at least 2 people, one therefore is forced to calculate the probability not only of 2 people but also $P(3)$, $P(4)$, . . . , $P(20)$ and to sum over all these terms. Another example might be the probability of having *less than* 2 defective parts in a sample of 10 with a given p. These questions can be answered by summing over the individual or point probabilities. In the first example, it would be easiest to calculate $P(0)$ and $P(1)$ and subtract their sum from 1. In the latter example, on the other hand, to get $P(r < 2)$ one should probably calculate $P(0)$ and $P(1)$ and then add them. Many binomial probability applications involve cumulative probabilities. Thus, most published binomial tables provide cumulative probabilities; individual or point probabilities are readily determined from cumulative tables. Appendix Table A.1 is such a table.

Look at Table A.1. Remember that the binomial distribution is a family of distributions depending on combinations of n and p. The table contains values of n from 1 through 20, 30, 40, 50, 75, and 100. For each n, probabilities for p values from .01 to .99 at increments of .01 are given. Since the table is cumulative, it deals with the *upper* end of the distribution for $p \leq .50$ and the *lower* end for $p > .50$. For example, with $n = 20$ and $p = .50$, the probability of 12 or more successes is .2517. For this illustration, $P(r \geq 12|n = 20, p = .50) = .2517$. The value .2517 is the sum of the point probabilities for all the values of r from 12 through 50.

Point probabilities are determined by subtracting two successive cumulative probabilities. For example, suppose one wishes to determine the probability of exactly 20 successes out of 50 trials with $p = .45$, i.e., $P(r = 20|n = 50, p = .45)$. Note that $P(r \geq 20) - P(r \geq 21)$ defines $P(r = 20)$. Thus, the point probability that r equals the integer 20 is .8026 − .7138, or .0888. In the above example, suppose one wanted to determine the probability that r is less than 20 given $n = 50$, $p = .45$. This involves subtracting $P(r \geq 20)$ from 1, i.e., $P(r < 20|n = 50, p = .45) = 1 - P(r \geq 20|n = 50, p = .45)$. In this case, the probability of r being less than 20 is $1 - .8026 = .1974$.

Use of the binomial table can be summarized by solving the inequalities and equalities presented in Table 6.4. These calculations, along with the explanation of each, are shown below:

1. $P(r = 8| n = 10, p = .50) = P(r \geq 8) - P(r \geq 9) = .0440$. Look at Table A.1 under $n = 10, p = .50$. Find the value where $r = 8$ (.0547). Subtract from this the value where $r = 9$ (.0107).

2. $P(r \geq 8|n = 10, p = .50) = .0547$. Read directly from Table A.1. Find $n = 10, p = .50, r = 8$.

3. $P(r > 8|n = 10, p = .50) = P(r \geq 9 |n = 10, p = .50) = .0107$. Since the probabilities in Table A.1 are cumulative, to find the probability of *r greater than* 8, one must find $n = 10, p = .50$, and $r = 9$ in the table.

4. $P(r \leq 8|n = 10, p = .50) = 1 - P(r \geq 9|n = 10, p = .50) = .9893$. To find $r \leq 8$, look in Table A.1 and find the value where $n = 10, p = .50$, and $r = 9$. Subtract this value (.0107) from 1.0000. This gives the probability of 8 or less successes in 10 trials.

5. $P(r < 8|n = 10, p = .50) = 1 - P(r \geq 8|n = 10, p = .50) = .9453$. To determine the probability of $r < 8$, find the value for $n = 10, p = .50$, and $r = 8$. Subtract this value (.0547) from 1.0000.

ILLUSTRATIVE USES OF THE BINOMIAL DISTRIBUTION

One of the best illustrations of the use of the binomial distribution is *Casteneda v. Partida,*[1] a case decided by the Supreme Court in 1977. Under the "key man" system for selecting grand juries in Texas, jury commissioners were appointed by a state district judge to select prospective jurors from different portions of the county. After the selection process was completed, the district judge proceeded to test their qualifications. In addition to being a citizen of the state and of the county in which he was to serve and a qualified voter, a grand juror had to be of sound mind and good moral character, be literate, have had no prior felony conviction, and be under no pending indictment or other accusation.

The respondent, a Mexican-American, was convicted of a crime in a Texas district court. After he had exhausted his state remedies on his claim of discrimination in the selection of the grand jury that had indicted him, he filed a habeas corpus petition in the federal district court, alleging a denial of due process and equal protection under the Fourteenth Amendment. He claimed gross underrepresentation of Mexican-Americans on the county grand juries. On the basis of the evidence before it, the district court concluded that the respondent had made out a weak prima facie case of invidious discrimination. The court concluded that its doubts about the reliability of population and grand jury statistics offered by the respondent from census and county records, coupled with its opinion that Mexican-Americans constituted a "governing majority" in the county, caused it further to conclude that the prima facie case was rebutted by the state. The petition was dismissed. The U.S. Court of Appeals for the Fifth Circuit reversed this decision, holding that the state had failed to rebut the respondent's prima facie case.

The Supreme Court affirmed the ruling of the court of appeals. The court pointed out that the statistical "disparity proved by the 1970 census statistics showed that the population of the county was 79.1% Mexican-American, but that, over an 11-year period, only 39% of the persons summoned for grand jury service were Mexican-American. . . . The mathematical disparities that have been accepted by this court as adequate for a prima facie case have all been within the range presented here."[2] Based on this evidence as well as the method of jury selection and other relevant testimony as to the manner in which the selection process was implemented, the court concluded that the proof offered by the respondent was sufficient to demonstrate a prima facie case of intentional discrimination in grand jury selection. To illustrate the degree of disparity, the court utilized the binomial distribution where only two outcomes—Caucasian or Mexican-American—are being analyzed:

> If the jurors were drawn randomly from the general population, then the number of Mexican-Americans in the sample could be modeled by a binomial distribution. . . . Given that 79.1% of the population is Mexican-American the expected number of Mexican-Americans among the 870 persons summoned to serve as grand jurors over the 11 year period is approximately 688. The observed number is 339. Of course, in any given drawing some fluctuation from the expected number is predicted. The important point, however, is that the statistical model shows that the results of a random drawing are likely to fall in the vicinity of the expected value. . . . The data for the 2½-year period during which the State District Judge supervised the selection process similarly support the inference that the exclusion of Mexican-Americans did not occur by chance. Of 220 persons called to serve as grand jurors, only 100 were Mexican-Americans. The expected Mexican-American representation is approximately 174 . . . as calculated from the binomial model.[3]

A number of relevant articles concerning application of the binomial distribution have been published in law journals. Shoben, writing in *Harvard Law Review*, states that "[t]he use of the binomial test is appropriate when evaluating the likelihood of a result composed of a series of events, each with only two possible outcomes, such as the selection of either a black or a white from the relevant population of school teachers."[4] Shoben points out that, in such a case, the binomial allows a comparison to be made between a racial composition of the sample selected and the known racial composition of the population. She further states:

> The use of the binomial test requires the assumption that the probability of selecting a member of either group remains unchanged throughout the selection process. This will be the case when selection is made from a theoretically infinite population or from a finite population with replacement. For example, the probability of selecting a woman from an infinite population which is 50% female will remain .5 with each selection of either a

male or female. When a population is very large, it can safely be assumed that the requirement is satisfied even though the population is finite.[5]

Shoben raises an interesting point in suggesting that "[t]he *Casteneda* decision was raised, in part, by a 1966 article in the *Harvard Law Review* in which Michael Finkelstein urged that statistical inferences based on the binomial distribution would assist the evaluation of claims that racial minorities are unconstitutionally excluded from grand juries."[6] Finkelstein's article is among the suggested references at the end of this chapter.

NOTES

1. 430 U.S. 482 (1977).
2. *Id* at 495–96.
3. *Id* at 496–97 n. 17.
4. Elaine W. Shoben, "Differential Pass-Fail Rates in Employment Testing: Statistical Proof Under Title VII," *Harvard Law Review* 91 (1978): 796.
5. *Ibid.*
6. *Ibid.*, 795.

SELECTED REFERENCES

Feller, W. *An Introduction to Probability Theory and Its Applications*, Chap. 6. 2d ed. New York: John Wiley & Sons, 1960.

Finkelstein, Michael O. "The Application of Statistical Decision Theory to the Jury Discrimination Cases." *Harvard Law Review*, Vol. 80 (1966), pp. 338–65.

Jedamus, Paul J. and Robert Frame. *Business Decision Theory*, Chap. 3. New York: McGraw-Hill Book Co., 1969.

Mendenhall, William and James E. Reinmuth. *Statistics for Management and Economics*, Chap. 6. Belmont, Calif.: Duxbury Press, 1971.

Mosteller, F., R.E.K. Rourke and G. B. Thomas, Jr. *Probability with Statistical Applications*, Chap. 7. Reading, Mass.: Addison-Wesley Publishing Co., 1961.

National Bureau of Standards. *Tables of Binomial Probability Distribution*. Washington, D.C.: Government Printing Office, 1949.

Spurr, W. and C. P. Bonini. *Statistical Analysis for Business Decisions*, Chap. 8. Homewood, Ill.: Richard D. Irwin, 1967.

CHAPTER 7

THE NORMAL DISTRIBUTION

As noted in Chapter 6, probability distributions of continuous random variables are referred to as continuous probability distributions. Probability as defined for a random variable refers to the probability that the variable assumes some value in a specified interval. The continuous probability distribution is pictured as a smooth curve; the area under the curve between two values represents the probability that the random variable assumes any value in that interval. Means and variances of continuous distributions are analogous to those measures as defined for discrete distributions.

Probably the best known and most useful of all probability distribution is the *normal distribution.* Also known as the *Gaussian curve* or the *normal curve of error,* the normal distribution is a continuous distribution. Because of its applicability to a wide range of populations, the discussion of continuous distributions will be limited to the normal distribution. The key importance of this distribution is derived from three sources:

1. The normal distribution provides a useful means of approximating many other distributions, e.g., the binomial distribution.
2. Many real world variables are normally or approximately normally distributed.
3. Most importantly, the normal distribution is extremely useful in sampling theory, i.e., a variety of useful sampling distributions conform to the normal distribution.

Study of the normal distribution began during the 1700s, when it was found that when physical characteristics such as weights, speeds, length, and volumes were measured repeatedly, variations occurred in the measurements. The variations occurred for several reasons. For example, the instruments with which the measurements were taken had built-in limits of accuracy. The manner in which different experimenters used the instruments varied. There were variations in the abilities of the experimenters to read the scales on the instruments accurately. Finally, conditions during which the measurements were taken were not always identical; variations in the measurements occurred in very similar patterns. Carl Friedrich Gauss, a 19th century astronomer who attempted to find the exact orbit of a planet, theorized that successive measurements yielding different values had a "normal" curve of error distributed around the true value. Hence, the normal distribution, a bell-shaped curve, was known for many years as the normal curve of error or the Gaussian curve. The normal distribution is illustrated in Figure 7.1.

DEFINITION OF THE NORMAL DISTRIBUTION

Mathematically the normal distribution is defined as:

$$P = \frac{1}{\sigma\sqrt{2\pi}} \cdot e^{-(x-\mu)^2/2\sigma^2} \qquad (7.1)$$

where

P = normal probability or the height of the curve for a given value of x
μ = mean of the distribution of x values
σ = standard deviation of the distribution
x = independent variable
π = 3.14159
e = 2.71828

Note that for any given value of x, P can be determined by knowing only the mean and standard deviation of the distribution. Thus, the normal distribution is defined by two parameters. Each combination of values of μ and σ defines a different distribution, although all of these distributions are normally distributed. The relationship of these distributions is perhaps more easily understood by considering an analogy with circles. Each value of r, the radius, defines a

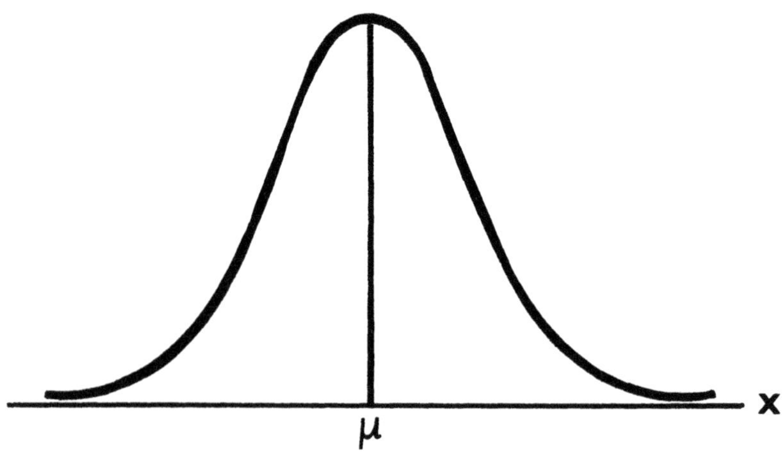

Figure 7.1
The Normal Distribution

different circle (the circumference is $2\pi r$ and the radius is πr^2), although each circle has the same shape.

A little investigation of Equation 7.1 will show that the normal curve tails off to plus and minus infinity and that its highest value occurs where $x = \mu$, at which point P becomes equal to $1/\sigma\sqrt{2\pi}$. Further investigation would prove that the two points of inflection on the curve occur at distances of $-\sigma$ and $+\sigma$ from μ.

Since much of inferential statistics rests upon the normal distribution, it is helpful to be aware of the characteristics of this distribution. The normal distribution has a probability density function which is smooth, symmetric, continuous, and bell-shaped. Consequently, the mean, median, and mode are identical. The area under the curve over any interval on the horizontal axis represents the probability of the random variable, x, taking on a value in that interval. The area under the curve sums to one.

As noted earlier, a normal distribution is completely determined by its mean and standard deviation. The curve reaches a maximum at its mean. One-half of the area lies on either side of the mean. The greater the value of σ, the greater the dispersion of the distribution.

Areas of the normal distribution are customarily related to standard deviations from the mean. For example, approximately 68.26 percent of the area under the normal curve lies within $\pm 1\sigma$, 95.44 percent lies within $\pm 2\sigma$, and 99.73 percent lies within $\pm 3\sigma$. Similarly, the interval between $\pm .67\sigma$ contains 50 percent of the area under the normal curve. This is illustrated in Figure 7.2.

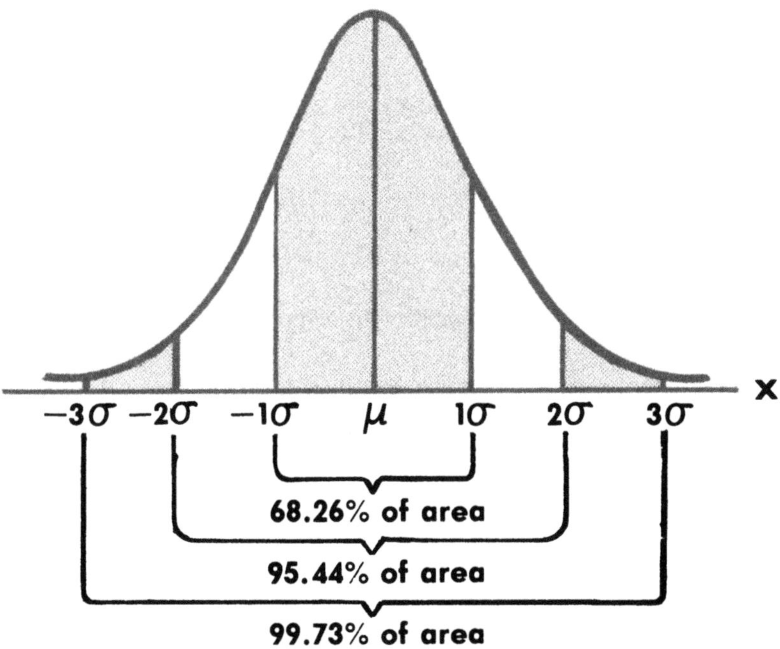

Figure 7.2
Areas under the Normal Curve

STANDARD NORMAL DEVIATE

To generate a normal distribution using Equation 7.1 would be a formidable task because there are an infinite number of possible values x could assume. Tables have been developed to facilitate construction of a normal distribution. For example, values for areas of the normal curve for selected ranges are found in Appendix Table A.2. Since a normal distribution is symmetrical, the table reflects values for only one side of the distribution. The area corresponding to $\mu \pm 1$ is found opposite $z = 1.0$. In Table A.2, this value is read as 0.3413. Since it corresponds to only one side of the distribution, the desired area is (2)(.3413) or .6826.

To use the table, various x values must be transformed into a standard measure called z. This is known as the *z value* but is often referred to as the *standard normal deviate* or *standard normal distribution*. The standard normal deviate has a mean of 0 and a standard deviation of 1.

To find probabilities for some random variable x with parameters μ and σ, it is necessary to transform these values of x to corresponding values of z, a variable which has a standard normal distribution. Thus, the probabilities that are determined for z are identical to the probabilities for x. Equation 7.2 can be used to transform values of x from a nonstandard distribution into standard z units:

$$z = \frac{x - \mu}{\sigma} \tag{7.2}$$

where

z = standardized normal variable
x = nonstandardized normal variable
μ = mean of the random variable x
σ = standard deviation of x

Consider the following example. A normally distributed random variable, x, has a mean of 50 and a standard deviation of 10, i.e., $\mu = 50$, $\sigma = 10$. Suppose one wishes to determine the probability that x has a value greater than 60—$P(x > 60)$.

The first step is to find a value of z that corresponds to $x = 60$. This is accomplished by the following transformation.

$$z = \frac{x - \mu}{\sigma}$$

Substituting the above values in Equation 7.2 yields a standardized z value:

$$z = \frac{60 - 50}{10} = \frac{10}{10} = 1$$

The probability that x is greater than 60 is equal to the probability that z is greater than 1. The latter value can be obtained from Table A.2. Simply stated, z is the distance between the mean of the distribution and the specified value of x measured in units of the standard deviation. The value $z = 1$ means that the value of x, 60 in this case, is one standard deviation above the mean.

To determine the probability that $x > 60$, simply find the probability that $z > 1$. From Table A.2, one can determine that the area between $z = 0$ and $z = 1$ is 0.3413 (Figure 7.3). Since the entire area to the right of the mean is equal to 0.5, the part of the area to the right of $z = 1$ must be $0.5 - 0.3413$, or 0.1587. Therefore, $P(x > 60) = 0.1587$.

The table of areas under the normal probability curve provides areas or probabilities for values of $z = 0$ to $z = 4.0$. The probabilities in the body of the table correspond to areas between the mean and a specified value of z. However, these probabilities are provided only for positive values of z. Because of the symmetry of the distribution, all areas below the mean which correspond to the negative values of z are identical to the corresponding values above the mean. Therefore, to find the area between the mean and some negative value of z, look up the area between the mean and the corresponding positive value.

Suppose one wanted to determine $P(x < 40)$. Since $z = \frac{x - \mu}{\sigma} = \frac{40 - 50}{10} = -1.0$, the probability that x is less than 40 is equal to $0.50000 - 0.3413$, or 0.1587—exactly the same as $P(x > 60)$. This is illustrated in Figure 7.4.

Now suppose the following values of x were given:

1. $P(30 < x < 50)$. This is the probability that x lies between 30 and 50, which is illustrated in Figure 7.5(a). The z value is -2.0. This is the same as $P(-2.0 < z < 0)$. The area under the normal curve lying between 0 and -2.0 is .4772; therefore, $P(30 < x < 50)$ is .4772.

2. $P(50 < x < 75)$. Illustrated by Figure 7.5(b), this refers to the probability that x falls between 50 and 75. Or, stated in terms of the standardized normal distribution, this is $P(0$

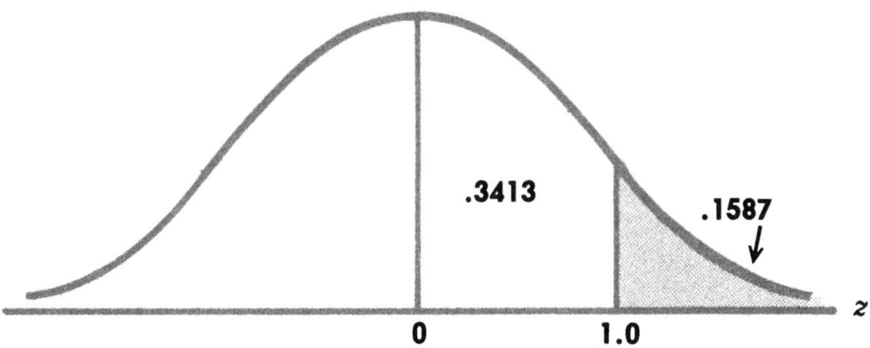

Figure 7.3
Probability of x Being Greater than 60

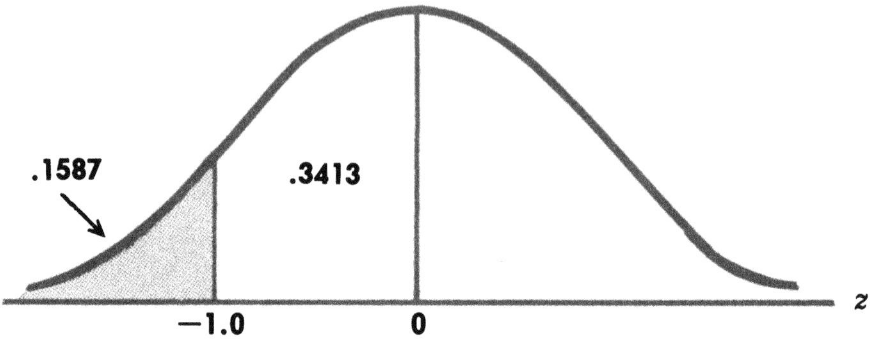

Figure 7.4
Probability of x Being Less than 40

$< z < 2.5$) or the probability that z falls between 0 and 2.5. The area under the normal curve corresponding to 2.5 standard deviations is .4938. Thus, the probability that x falls between 50 and 75 is .4938.

3. $P(40 < x < 80)$. This is the area under the curve between the values 40 and 80. In terms of the standard normal distribution, this corresponds to z values of -1.0 and 3.0 as calculated by Equation 7.2. This is the same as $P(-1.0 < z < 0) + P(0 < z < 3)$ or .3413 + .49865 = .83995. This is shown in Figure 7.5(c).

4. $P(60 < x < 80)$. Here one wishes to know the probability of x lying between the values of 60 and 80 when $\mu = 50$. Converting this to the standardized normal distribution,

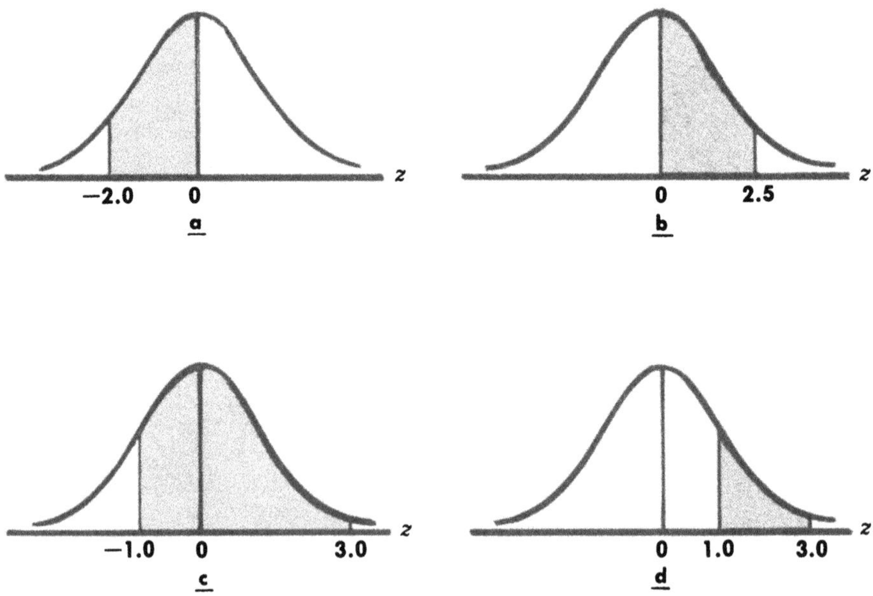

Figure 7.5
Illustrations Using the Standardized Normal Distribution

the relevant z values become 1.0 and 3.0, i.e., $P(1.0 < z < 3.0)$. In this instance, the area to be considered falls to the right of the mean, μ. Two separate calculations are involved. First, the probability that z lies between 0 and 1.0 must be determined; this is equal to .3413. Second, the probability of z being between 0 and 3.0 must be determined; this is equal to .49865. As shown in Figure 7.5(d), the difference in these two areas—.49865 − .3413—is the probability of the value of the random variable x falling between 60 and 80 or z being between 1.0 and 3.0. This probability is .15735.

NORMAL APPROXIMATION TO THE BINOMIAL DISTRIBUTION

In statistics, one often encounters the situation in which the binomial distribution can be assumed but n, the number of trials, is so large that the calculations become tedious and a complete table of binomial values is not available for that n. In these situations, the normal distribution can, within reasonable ranges, be used to find a close approximation to the true binomial distribution. As a general rule, the normal approximation to the binomial may be used when n is large and p, the probability of success on any given trial, is *not* small. Most statisticians recommend that *both* np and $n(1 - p)$ should be greater than 5 before the normal approximation to the binomial can be applied.

To illustrate the approximation process, assume a sample of 100 persons is randomly selected for which p, the probability of selecting a male, is .40. The probability of selecting exactly 35 males—$P(r = 35|n = 100, p = .40)$—is found in Table A.1 by finding values where $r = 35$ and subtracting from this value where $r = 36$. Thus, the probability of selecting exactly 35 males is .0492.

When using the normal distribution to approximate the binomial, a continuous distribution is employed to represent the histogram of a discrete variable, and the discrete variable is treated as if it were continuous. A bar of the histogram of a discrete variable corresponds to a single value of the variable. Since only integer values are defined for this type of variable, the bar is assumed to extend halfway to the integer preceding and the integer following the central value. The extension of the histogram in this manner is called the *continuity correction factor;* as a general rule, the continuity correction factor can be ignored whenever n is greater than 100. Areas under the normal density function are found in terms of deviations from the mean; a similar approach is used to find normal areas to approximate binomial probabilities. Figure 7.6 illustrates the approximation of the binomial distribution with a normal distribution.

If x is a random variable with a binomial distribution, approximate probabilities can be obtained from the normal distribution with $\mu = E(x) = np$ and $\sigma = \sqrt{np(1 - p)}$. For the example cited earlier,

$$E(x) = np = (100)(.40) = 40.00$$

and

$$\sigma = \sqrt{np(1 - p)} = \sqrt{(100)(.40)(.60)} = 4.90$$

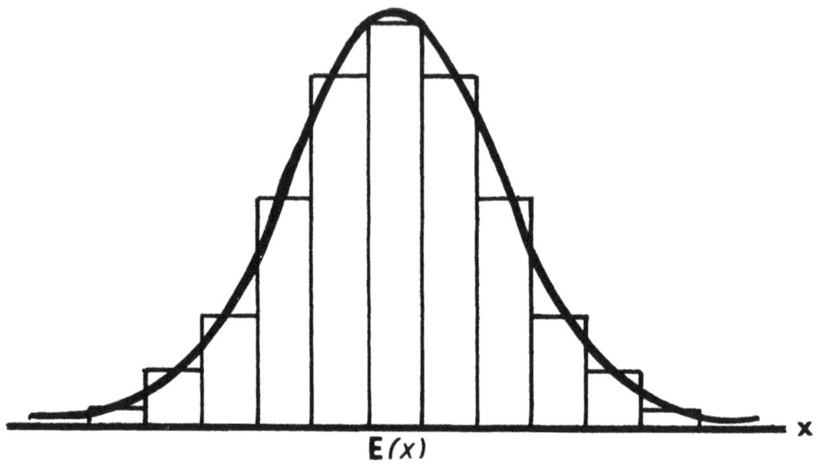

Figure 7.6
Binomial Distribution Approximated by a Normal Curve

Since the histogram at $x = 35$ extends from 34.5 to 35.5, the normal approximation is found by expressing these values as deviations from $E(x)$ in terms of standard deviations. This requires calculation of the standardized z value:

$$z = \frac{34.5 - 40.00}{4.90} = -1.12$$

$$z = \frac{35.5 - 40.00}{4.90} = -.92$$

These values of z are read in Table A.2 as 0.3686 and 0.3212, respectively. The normal approximation of this binomial probability is $0.3686 - 0.3212 = .0474$, which compares favorably with the exact probability of 0.04920 read from Table A.1.

Now that the validity of the normal approximation to the binomial has been illustrated, a more sophisticated example for which binomial tables are not generally available will be used. Assume a survey indicates that 70 percent of the readers of a daily newspaper read the classified section of the paper. To verify this finding, the marketing department of the newspaper randomly selects 200 readers from its master file of subscribers and interviews each to determine if he or she reads the classified section of the paper. If the survey findings are correct, what is the probability that at least 150 of the shoppers in the sample are readers of the classified section?

Let x represent the random variable associated with the number of readers. Thus, x has a binomial distribution with $n = 200$ and $p = .70$. $E(x) = np + 140$, and $\sigma = \sqrt{(200)(.70)(.30)} = 6.48$. Since $np = 140$ and $n(1 - p) = 60$, both of which are greater than 5, the normal approximation can be used. Since the binomial mean and standard deviation are 140 and 6.48, respectively, a normal distribution having the same mean and standard deviation can be used. Because n exceeds 100, the continuity correction factor can be ignored. To find $P(x \geq 150)$:

$$z = \frac{150 - 140}{6.48} = \frac{10}{6.48} = 1.54$$

Thus, $P(z \geq 1.54) = 0.50000 - 0.4382 = .0618$.

ILLUSTRATIVE USES OF THE NORMAL DISTRIBUTION

In a number of cases, the courts have either explicitly or implicitly assumed a normal distribution in utilizing statistical data. One case in which the normal curve is named as such is *United States v. Georgia Power Company,*[1] a 1973 employment discrimination case before the court of appeals. The decision of the district court was affirmed in part, vacated in part, and remanded.

A key portion of the case revolved around the use of employment tests by Georgia Power, a practice which the plaintiff contended was discriminatory. Georgia Power acknowledged its burden to validate the tests it had used and attempted to do so with expert opinion evidence. This took the form of a "post testing study made by Dr. Loren Hite which concluded that, after each of the tests had been properly weighted for the jobs involved, the resulting scores bore a positive relationship to job performance ratings by supervisors."[2]

The court noted that since Dr. Hite's study and his testimony were the only specific proof offered to uphold Georgia Power's tests, "the district court necessarily based its validation findings on this evidence alone."[3] In denying the validity of Dr. Hite's approaches and conclusions, the court noted:

> In his testimony Dr. Hite complained of the lack of a representative sample of applicants in the data with which he worked. Completely aside from the virtual absence of blacks in the sample, he noted that the absence of any evaluation of those who failed the tests caused statistical difficulties because he was cut back to the use of distributions which had about one-half the range of test manual norms. The lower 40% of the predicted bell-shaped curve of applicants (which lower range included the majority of the blacks tested) was never evaluated. Rather than attempting to supply this deficiency through ad hoc evaluation of probationary employees or the use of validity studies from other organizations with comparable jobs . . . Dr. Hite proceeded . . . despite the loss of the group.[4]

NOTES

1. 474 F.2d 906 (1973).
2. *Id* at 912.
3. *Id* at 913.
4. *Id* at 916.

SELECTED REFERENCES

Benson, Purnell, H. "Fitting and Analyzing Distribution Curves of Consumer Choices." *Journal of Advertising Research,* Vol. 5, No. 1 (March 1965), pp. 28–34.

Bryant, Edward C. *Statistical Analysis.* Revised ed. New York: McGraw-Hill Book Co., 1965.

Chew, Victor. "Some Useful Alternatives to Normal Distribution." *The American Statistician,* Vol. 22 (June 1968), pp. 22–24.

Guenther, William C. *Concepts of Statistical Inference,* Chap. 2. 2d ed. New York: McGraw-Hill Book Co., 1973.

Miller, Irwin and John E. Freund. *Probability and Statistics for Engineers,* Chaps. 3, 4. Englewood Cliffs, N.J.: Prentice-Hall, 1965.

Yamane, Taro. *Statistics: An Introductory Analysis,* Chap. 6. 3d ed. New York: Harper & Row, 1973.

CHAPTER 8

STATISTICAL DECISION THEORY

The process of decision making, a concept familiar to individuals and groups, involves choosing among alternative courses of action on the basis of what is known or thought to be known about the consequences of each decision option. For unless there are alternatives from which to choose, no decision problem exists. The rational decision maker should attempt to select the action or actions which will produce the "best" result or consequence in the context of his decision problem.

In recent years, classical statistics has been augmented by a more comprehensive approach called *statistical decision theory*. These additions to classical theory—sometimes called "Bayesian decision theory"—encompass the study of decisions involving risk and uncertainty; monetary or utility values are implicitly incorporated into the determination of the best decision.

This chapter contains a brief presentation of some of the principal concepts of decision theory that have relevance for practicing attorneys. Recent legal literature has been replete with references to potential applications of decision theory. Nevertheless, the reader is cautioned that this chapter is not to be considered an exhaustive or definitive presentation. It is intended rather to acquaint readers with Bayesian objectives. Those interested in a more vigorous pursuit of the subject should become familiar with the references provided at the end of the chapter.

ELEMENTS OF DECISION THEORY

Before discussing applications of decision theory, certain basic concepts must be defined. These concepts or elements are required to provide a general description of decision problems as well as a framework for obtaining solutions.

Uncertainty and Risk

Uncertainty in the decision-making context, a rather common state, occurs in those situations in which no prior information is available concerning the likelihood of certain events. In other cases, it is possible to obtain prior information about the probabilities of occurrence of events. Decisions made with the benefit of such information, whether obtained objectively from data analysis or subjectively from feelings or experience, are made under a condition called *risk*. Where complete information is available about which event or events will occur, a decision can be made under conditions of *certainty*.

The discussion in this chapter focuses on decisions made under *risk*, i.e., those conditions in which probabilities can be attached to the occurrence of events affecting decision making.

Actions, States of Nature, and Probabilities

In decision theory analysis, *actions* are the alternatives available to the decision maker. They are the factors over which he has control. Assume, for example, a law firm is considering expanding into one of three locations. That is, it may move to one of three cities or it may decide not to move. The choice of these three locations plus the possible decision not to move represent the actions available to the law firm in the context of decision theory. In decision theory notation, actions are denoted A_i; the options available to the firm are represented by A_1, A_2, A_3, and A_4.

Those things over which the decision maker has little or no control, called *states of nature or events*, comprise another element of decision theory. States of nature in the above example are these: depending on the state of the economy, (1) gross national product (GNP) in the national economy may increase by 20 percent; (2) it may increase by 10 percent; or (3) it may, in the worst possible case, decline by 5 percent. The law firm as the decision maker in this case has no control over the states of nature, i.e., the status of the economy. States of nature are represented by S_j. In this example, the three states are denoted S_1, S_2, and S_3.

Probabilities, another element in decision theory, are the chances of the occurrence of the various states of nature. These probabilities can be empirical or subjective in nature. Since the states of nature are mutually exclusive and collectively exhaustive, the sum of the probabilities must equal 1.00. With assistance from economic consultants, the law firm may subjectively assign probabilities to the various states of nature. Probabilities are denoted $P(S_j)$. In this case, they are $P(S_1)$, $P(S_2)$, and $P(S_3)$.

Payoff Tables and Conditional Values

The three elements discussed previously combine to form a *payoff table* or *matrix*. Associated with each combination of an action and a state of nature is a payoff, a consequence or outcome that is usually expressed in monetary terms (gain or loss). A payoff is also called a *conditional value*, since it is the value corresponding to an action, given the occurrence of a particular state of nature. Payoffs or conditional values are usually represented by the Greek letter theta, θ. The set of actions, the set of states of nature, and payoffs or conditional values associated with each action and each state are illustrated in Table 8.1.

Table 8.1
General Format of Decision Theory Model

Actions \ States of Nature	$P(S_1)$ S_1	**Probabilities** $P(S_2)$ . . . S_2 . . .	$P(S_j)$ S_j
A_1	θ_{11}	θ_{12} . . .	θ_{1_j}
A_2	θ_{21}	θ_{22} . . .	θ_{2_j}
.	.	.	.
.	.	.	.
.	.	.	.
A_i	θ_{i_1}	θ_{i_2} . . .	θ_{i_j}

The problem posed earlier can now be illustrated in terms of a payoff matrix. Given the four actions available to the law firm, the three possible states of nature, the subjective probabilities of the occurrence of the various states, and the payoffs associated with each combination of action and state of nature, a payoff table can be constructed. The problem can be summarized as in Table 8.2.

The table may be explained as follows. If S_1, a GNP increase of 20 percent, occurs, profits to the law firm will increase by $400,000 if the firm expands to city A, by $450,000 if it goes into city B, by $375,000 if the firm sets up another office in city C, and by only $50,000 if the firm does not expand into any one of the cities. Conversely, if the economy falters and GNP declines by 5 percent, profits will decline by $50,000 if expansion to city A occurs, by $56,250 if expansion occurs in city B, by $46,875 if expansion takes place in city C, but by only $6,250 if no expansion occurs. In order to decide which act is optimal under the conditions of risk about the state of nature, some criterion of optimality is required. Without such a criterion, the payoff table serves principally as a device for displaying the states of nature, probabilities, actions, and payoffs or consequences in a concise and clear manner.

Table 8.2
Payoff Table, Law Firm Example, January 1983

Actions \ States of Nature	.15 *GNP Increases 20% (S_1)*	**Probabilities** .35 *GNP Increases 10% (S_2)*	.50 *GNP Declines 5% (S_3)*
A_1, Expand to city A	$400,000	$160,000	$−50,000
A_2, Expand to city B	450,000	180,000	−56,250
A_3, Expand to city C	375,000	150,000	−46,875
A_4, Do not expand	50,000	20,000	−6,250

Expected Monetary Value

There are several criteria for selecting an action that is optimal in some sense from among the various alternative actions. Only one of these, Bayes' criterion, will be considered here. Bayes' criterion is an expected value criterion computed from a set of probabilities, in this case the probabilities associated with the various states of nature.

Bayes' criterion states that the expected payoff should be computed for each action; that action for which the expected payoff is best should be selected. "Best" refers to the highest value if the payoffs are expressed in terms of profit or gain; it refers to the lowest value when payoffs are expressed in terms of cost or loss.

Since Bayes' criterion is an expected value criterion, an expected value must be computed for each action. The expected value of an action is simply a weighted average taken over all values of the states of nature where the weights are the probabilities. Since payoffs in the problem posed earlier in this chapter, as well as in most business problems, are expressed in terms of dollars or monetary value, the expected value of an action is called the *expected monetary value,* abbreviated *EMV*. The expected monetary value of an action is

$$EMV(A_i) = \sum_j P(S_j)(\theta_{ij}) \qquad (8.1)$$

Equation 8.1 simply means that, in order to obtain the *EMV* of an act, multiply each conditional payoff of that action for a given state of nature by the probability of that state and sum all these products over all the states of nature.

Table 8.3 repeats the data provided in Table 8.2. The expected monetary value of each action is provided. According to Bayes' criterion, the optimal action is A_2, which involves expansion to city B. The *EMV* of this act, $102,375, is higher than the *EMV* of any other action.

Table 8.3
Payoff Table with Expected Monetary Value, Law Firm Example, January 1983

States / Actions	Probabilities .15	.35	.50	
	S_1	S_2	S_3	*EMV*
A_1	$400,000	$160,000	$−50,000	$ 91,000
A_2	450,000	180,000	−56,250	102,375
A_3	375,000	150,000	−46,875	85,312
A_4	50,000	20,000	−6,250	11,375

Opportunity Loss

In decision theory applications, it is common to specify the conditional values of an action for any given state of nature in terms of *opportunity loss* instead of payoff. Opportunity loss is the absolute value of the difference between the payoff actually received for the action selected and the payoff which could have been obtained had the best or optimal action been selected. The opportunity loss associated with the optimal action for a given state of nature is always 0, i.e., no opportunity loss. For any other action given the state of nature, the opportunity loss is always a positive value. If all the conditional values are specified in terms of opportunity loss, then the optimal action is always the action with the lowest expected opportunity loss, *EOL*.

An opportunity loss table results if all conditional values are shown in terms of opportunity loss. To convert a payoff table to an opportunity loss table, find the optimal action for each state of nature—the action with the highest payoff. The opportunity loss for that action and the given state of nature is 0. For all other actions and the given state of nature, the opportunity loss is the absolute difference between the conditional value of that action and the conditional value of the optimal action. The opportunity loss table corresponding to Table 8.3 is shown in Table 8.4.

To understand how the opportunity losses in Table 8.4 were calculated, consider S_1, an increase in GNP of 20 percent. Given this state of nature, the optimal action is A_2, expand to city B. The firm's payoff for that action and state of nature is \$450,000 in increased profits; the opportunity loss corresponding to the best action is 0—there is no opportunity loss. If the firm had selected A_3—i.e., expanded to city C—the opportunity loss would have been \$75,000, the difference between the value of the optimal action and A_3. In a like manner, the optimal action for S_3 is A_4, do not expand. The firm's payoff for that action and state of nature would be a loss of \$6,250; the opportunity loss is 0. If, for example, the firm had selected A_1, the opportunity loss would have been

Table 8.4
Opportunity Loss Table with Expected Opportunity Loss, Law Firm Example, January 1983

States / Actions	Probabilities .15	.35	.50	
	S_1	S_2	S_3	*EOL*
A_1	\$ 50,000	\$ 20,000	\$43,750	\$ 36,375
A_2	0	0	50,000	25,000
A_3	75,000	30,000	40,625	42,063
A_4	400,000	160,000	0	116,000

\$43,750—the absolute difference between the value of the optimal act of \$−6,250 and value of the action chosen, \$−50,000. The same procedure is followed for all other actions for each state of nature. The expected opportunity loss for each action, *EOL,* is calculated in the same way as the *EMV*.

The optimal action is that action with the lowest expected opportunity loss. This is A_2, with an *EOL* of \$25,000. From Table 8.3, it will be seen that this is the action which has the highest *EMV*. It should be noted that the optimal action will be the same whether the choice is based on *EMV* or *EOL*.

Expected Value of Perfect Information

A logical extension of the payoff and opportunity loss tables is the *expected value of perfect information* or *EVPI*. *EVPI* can be used as a rough analysis of whether it is worthwhile to spend money for further information. In other words, *EVPI* is a measure used to evaluate the worth of perfect information.

If a decision maker had perfect knowledge about the state of nature, then he would always be able to choose that action which represents the best or optimal decision under the circumstances. The difference between his payoff with perfect information—i.e., under conditions of certainty—and his expected payoff under uncertainty represents the *cost of uncertainty*. This measure is the expected opportunity loss of the *optimal* action under the Bayes' criterion. With perfect information, opportunity loss would always be 0; under those conditions, the right decisions would always be made. Thus, the expected opportunity loss of the optimal act under uncertainty represents the amount by which the decision maker could improve the consequences of his decision if he had perfect information. *EVPI* is defined as being exactly equal to the expected opportunity loss of the optimal act. In the previous example, *EVPI* was found to be \$25,000. This figure represents the maximum amount the firm could pay to purchase perfect information. Since it is known that perfect information can never be delivered, the maximum expenditure for additional information should be less than \$25,000.

ILLUSTRATIVE USES OF DECISION THEORY

Although recent legal literature makes reference to a number of potential uses of decision theory, most of the references are academic in nature. There are, however, some practical uses of decision theory. The following case illustrates use of the concept.

Donald Reuben Brantley, a teenager, was injured in a truck–motorcycle accident, resulting in the amputation of one of his legs. The accident occurred when the driver of the truck ran through a stop sign and struck the motorcycle on which Mr. Brantley was riding. The accident impaired his lifetime earning capacity by restricting the types and amount of work he will be able to perform in the future. Because of this, his parents brought suit against the driver of the truck. Only 15 years old at the time of the accident, Mr. Brantley was assumed to have a normal life expectancy and to remain employed until the normal retirement age of 65.

The difference between his expected earnings to age 65 prior to the accident and his expected lifetime earnings after the accident constitutes his lost earnings.

Three basic problems had to be considered in projecting the plaintiff's lost earnings. To a large extent they were interrelated. First, Mr. Brantley's career interests were not known. It is difficult for the average person to be able to determine his vocational interest at 15 years of age. A second factor complicating the analysis was the fact that Mr. Brantley had not entered the labor force and thus had no record of past earning. Third, the psychological impact of his injury on his vocational motivation and earnings capacity could not be incorporated into the analysis.

As the result of the facts and uncertainties surrounding this case, especially the absence of definitive vocational aptitude, a general approach had to be used in preparing initial estimates. The procedure adopted was to delineate some probable vocational areas that were precluded by the accident; estimated lifetime earnings in these vocations were compared with those that Mr. Brantley might now be able to receive as a result of his impaired earnings ability. The vocational areas were developed with the assistance of vocational and guidance counselors.

It was assumed that, because of a combination of circumstances, Mr. Brantley would terminate his formal education with a high school diploma. Most jobs for which he would qualify would be those requiring limited mobility, probably of a sedentary nature. It is a well-known fact that a person with a high school education, below average earning ability, and limited mobility will normally have a difficult time finding a high-paying job. Many of the higher-paying jobs for which he would qualify are those connected with construction and manufacturing, both of which were precluded by his impairment. His injury would also prevent his choosing a military career as well as other areas requiring mobility.

The data in Table 8.5 project the estimated loss in earning incurred by Mr. Brantley as a result of the accident. These figures represent the present value of lost earnings among the various occupational groups Mr. Brantley might have chosen prior to the accident and those occupations he may be forced into subsequent to the accident. For example, if he had selected the construction field as an

Table 8.5
Estimated Present Value of Lost Earnings
among Various Occupational Groups, Daniel Reuben Brantley

	Precluded Occupations				
Possible Occupations	*Construction* *(p = .25)*	*Equipment Operator* *(p = .25)*	*Mechanic* *(p = .25)*	*Military* *(p = .25)*	**Expected Value**
Clerical	$439,238	$361,018	$157,224	$ 17,947	$243,857
Sales	465,325	387,105	183,311	44,034	269,944
Inspector	405,770	327,550	123,756	−15,521	210,389
Teacher's aide	408,152	329,932	126,138	−13,139	212,771

occupational choice prior to the accident but is now forced to accept employment as an inventory or shipping clerk, the present value of his loss in earnings over his lifetime is $439,238.

Viewing Table 8.5 in statistical decision theory context, the "possible occupations" rows represent the actions available to Mr. Brantley, while the "precluded occupations" columns are the states of nature. Since the probability of his being employed in one of the occupational groups precluded by the accident is unknown, it was assumed that each of the precluded occupations (states of nature) was equally likely to have occurred, i.e., the probability of each occurring is .25. Using Bayes' criterion, the expected values for each of the possible occupations were computed. If, for example, Mr. Brantley were to decide to become a teacher's aide, the expected monetary value of this action, expressed as the present value of the loss in earnings associated with his accident, would be $212,771. In a like manner, the expected value of his lost lifetime earnings should he become a clerical worker was found to be $243,857.

In light of the circumstances and uncertainties surrounding this case, the decision theory analysis described here played a significant role in bringing about a favorable out-of-court settlement for the plaintiff.

SELECTED REFERENCES

Birmingham, Robert. "Remarks on 'Probability' in Law: Mostly, a Casenote and Book Review." *University of Georgia Law Review*, Vol. 12 (1978), pp. 535–51.

Hamburg, Morris. "Bayesian Decision Theory and Statistical Quality Control." *Industrial Quality Control*, Vol. 19, No. 6 (December 1962), pp. 10–14.

Raiffa, Howard. *Decision Analysis: Introductory Lectures on Choice Under Uncertainty*. Reading, Mass.: Addison-Wesley Publishing Co., 1968.

———, and Robert Schlaifer. *Applied Statistical Decision Theory*. Boston: Harvard University, Graduate School of Business Administration, Division of Research, 1961.

Savage, Leonard J. "Bayesian Statistics." In *Recent Developments in Information and Decision Processes*, edited by Robert E. Machol and Paul Gray. New York: Macmillan Co., 1962, pp. 161–94.

Schlaifer, Robert. *Probability and Statistics for Business Decisions*. New York: McGraw-Hill Book Co., 1959.

Weeks, Ben. "Bayesian Probability: A Quantitative Technique for the Practicing Attorney." *American Business Law Journal*, Vol. 13 (1975), pp. 83–102.

PART III

Analyzing Statistical Data

CHAPTER 9

SAMPLING AND SAMPLING DISTRIBUTIONS

Sampling is the process of selecting a sample from a population. Information obtained from samples is the basis for all statistical inference. A sample is some part or subset of a population. When only a part of a population can be observed, all conclusions about the population must be inferred from the sample data.

The objective of sampling is the selection of a representative group of items that constitute a small-scale replica of the population from which they are drawn. Since such a sample tends to reflect the properties of the population, statistics computed from the sample data provide estimates of the respective parameters. The larger the sample size, the closer a sample statistic tends to approximate a population parameter.

Samples are taken when it is either impractical or impossible to make observations on the entire population. Observing an entire population is impractical when the population is very large; such a procedure would be too expensive in terms of both time and money.

Inferences based on sample data will always include some error, referred to as sampling error. The likelihood that the sampling error exceeds any specified magnitude must always be specified in terms of a probability value. This is the reason a study of probability and probability distributions precedes the study of statistical inference.

Because populations are variable, a representative sample must be large enough to exhibit the same—or a similar amount of—variation as the population. The greater the variation in the population, the larger the sample must be to achieve a given degree of precision. Problems in sampling relating to the representativeness of the sample fall into two categories—sample size and sample selection. Sample size is discussed in Chapter 10; sample selection is addressed in this chapter.

PROBABILITY OR RANDOM SAMPLING

There is no "best" method of selecting a sample. As mentioned earlier, however, samples must be drawn to reflect the special conditions of the population and under certain rules in order to use the sample results for statistical inference. Samples drawn under these conditions are called *probability samples.* Major probability sampling techniques are simple random sampling, systematic sampling, stratified random sampling, and cluster sampling.

Simple Random Sampling

Probability sampling is alternately called *simple random sampling.* A sample from a finite population can be considered a simple random sample if all the items in the population have an equal chance of being included in the sample. Whenever the word *sample* is referred to in this book, it will mean a simple random sample unless otherwise specified.

The concept of simple random sampling may be illustrated by the selective service draft just prior to the outbreak of World War II. All draft-age males in the United States were required to register at local draft boards; each received a card with a number on it. To ensure an orderly draft, eliminate the possibility of local draft board bias, and assure that each male had an equal chance of being selected for military service, all the numbers were placed in a fish bowl. The secretary of war picked a number from the bowl, and all men with that number were required to report to their local draft boards. Successive drawing of numbers at random from the bowl filled draft quotas. Selection of potential jurors from registered voter lists is a similar illustration.

The type of simple random sampling noted above is sampling without replacement. Sampling can occur with or without replacement. Sampling with replacement means that, when each sample element is selected, it is returned to the population before the next selection is made. In replacement sampling, it is possible for the same elements to be selected for the sample more than once. In actual practice, replacement sampling is seldom, if ever, used.

One way to select a simple random sample is to assign numbers to each element in the population and then make a random selection of these numbers. This is essentially the procedure followed by the selective service system in World War II. To illustrate, assume a population of five elements: *a, b, c, d,* and *e.* The elements would be numbered 1, 2, 3, 4, and 5, respectively. On the first selection, each element has a probability of being selected of 1/5. On the second selection, each of the remaining elements has a probability of being selected of 1/4, while on the third selection, each would have a probability of 1/3. Since on each selection all the remaining elements have the same probability of being selected, a simple random sample results.

In actual practice, the way to assure that every item in the population has an equal chance of being chosen is to use a *table of random numbers.* An abbreviated portion of such a table is included here as Table 9.1. All the digits in a table of random numbers have been generated by some random computer process. In Table 9.1, the digits are grouped into sets of three. Any series of digits read across the rows or down the columns is random.

It is necessary to have a numbered list of the elements in the population in order to use this table. The list is called a *sampling frame.* Numbers are assigned

Table 9.1
Table of Random Numbers

952	520	422	479	017
115	645	655	157	842
548	157	523	505	640
436	571	595	270	119
777	325	559	344	133
673	905	494	246	295
350	288	307	898	197
359	517	236	390	265
225	727	406	117	722
995	252	938	476	375

to the elements in the frame. Thus, a random selection of numbers from the table results in a random selection of the corresponding elements that constitute the sample.

Suppose a law firm retains a statistician to select a random sample of telephone subscribers from the telephone directory of a large city in order to obtain information pertaining to pretrial publicity for a client. Assume the statistician uses a table such as Table 9.1. He can pick any starting point and read down or across in groups of two sets of digits. Suppose he starts in the upper left corner and reads down two sets. The first set of three digits indicates a randomly selected page; the second set corresponds to the name on the sample page. Reading from the upper left corner, the number 952 indicates the name of the first subscriber to be sampled appears on page 952 of the directory and it is the 115th name on that page. The second set of random numbers is 548436. The name of the second subscriber to be sampled appears on page 548, and it is the 436th name on that page. The remainder of the sample can be obtained in a similar manner by continuing to read down each column until the desired number of names has been obtained. Should a directory have fewer pages than the three-digit random number obtained from the table, omit that number, proceed down to the next random number, and obtain another series of digits. This procedure also applies to situations where a three-digit random number is larger than the number of names on a given page.

A newer procedure to provide the same result for telephone interviewing is the use of a computer to generate random telephone numbers. The primary advantage of this approach, other than reducing the time required to search a telephone directory, is the fact it provides access to unlisted as well as newly listed numbers. Neither of these types of telephone numbers would be found in a directory.

Systematic Sampling

When the members of a population are already physically arranged in some order, a random sample is frequently obtained without preparing a separate list or frame. Assume that the population consists of 3,200 invoices in a file drawer. If an auditor wanted to select 10 percent of these invoices for examination using a simple random sample, then each invoice would have to be numbered from 0001 to 3,200. In addition, numbers would have to be selected from a table of random numbers. The invoice to match each of the random numbers would have to be located in the file. As can be seen, a simple random sampling method would be time-consuming.

Rather than using the simple random sample, a *systematic* sampling procedure would be more appropriate in the above example. A systematic sample may be used if the population is arranged in an orderly manner. In the above example, a random sample can be obtained by selecting every Kth element as long as the values of the characteristics are unrelated to the order in which the elements are arranged, *provided the starting point is determined by some random method.* The value K is called the *sampling interval.*

If the auditor wanted to sample 10 percent of the invoices noted previously, the sampling interval is 10. (Alternately, suppose he had wanted to select 320 invoices, then the sampling interval would have been $K = N/n = 3{,}200/320 = 10$). The sample is obtained by selecting every 10th invoice in the file. The starting point for the sample is determined by randomly selecting a number between 1 and 10 by any random method of selection. In this case, suppose the number 5 is selected. Beginning with the 5th invoice in the file, every 10th invoice is selected until the sample of 320 is completed.

Stratified Random Sampling

Another commonly used sampling procedure is *stratified random sampling*. In this procedure, the population is divided into subpopulations called strata. Each stratum is sampled independent of other strata; a random sample, called a subsample, is drawn from each stratum. An attempt is made to define the strata so that the elements within a given stratum are relatively homogeneous compared to the population as a whole. Once all the subsamples are taken, they are combined to form the overall sample.

A stratified random sample is a refinement of probability sampling. The method of stratification still assures each item in the population a chance of being selected in the sample. For example, a stratified random sample of business firms might be stratified by number of employees, gross sales, type of business, or any other criteria considered to be of importance to the representativeness of the sample. Likewise, the population of a city might be stratified by age, income, or other variables.

Cluster Sampling

Cluster sampling involves the random selection of entire groups, called *clusters,* from the population. Observations may then be made on every single member of the cluster, in which case the procedure is called single-stage sampling, or a subsample may be taken from each cluster, resulting in two-stage sampling. The technique is frequently used in surveys of people requiring personal interviews. When this type of sampling is to be done in geographic areas, it is often called *area sampling*.

The Gallup Poll—as well as other nationally known surveys—uses basically the cluster sampling technique. In an interview with an editor of *U.S. News and World Report* several years ago, Dr. George Gallup provided the following information.

Q. Speaking of sampling: How big a sample do you take in a nationwide pool—how many people do you question?

A. We do it with a minimum of 1,500. This is always a source of contention and amazement. But this whole system is based on the laws of probability.

Q. How did you arrive at 1,500?

A. Empirically—simply through the years. We started with larger samples. There is an interesting thing about the laws of probability. This is that the size of the "universe," or total population group to be sampled, makes no difference at all in the necessary size of the sample. By this I mean that when polling the United States of America we don't need any bigger samples than if we are polling New York City, or Trenton, N.J.

Q. But, in using the law of probability, shouldn't you have a random sample?

A. Yes, and we do. What we've had to do is to break the country down by election districts—precincts. And we've picked those precincts, literally, out of a hat, so to speak. We just line them up and pick every *x* number of precincts, taking the various sizes into account. We do this right across the country. So we end up with a random selection of geographical areas.

Now, having chosen the districts at random, we send an interviewer in and he follows a set pattern. He can't just go and pick people out himself. We actually give him a block assignment. He may be instructed to start with the third house from the northeast corner and then go to other dwelling units that assure a random selection. The selection of those interviewed is out of his hands.

Q. Since you pick the precincts at random, isn't there a risk that you'll get a sample that is weighted to a Democratic neighborhood or to a Republican neighborhood?

A. No. This is all part of the statistical process. You see, we have the past voting record in all these districts. So what we do is stratify within the sample, to make sure that we have the right percentage of people from New England, say, or from any area of the country.

Q. Dr. Gallup, how do you answer the question we hear so often: "Well, they've never polled me?"

A. Actually, there's a very simple answer. . . . Now, if we polled 1,500 every week—and we don't send out a ballot every week—it would take over 1,400 years, assuming a stable population. The chances of being interviewed are, as you can see, in a population of 110 million adults . . . are like a drop in a bucket.[1]

Most probability sampling methods are variations or combinations of the methods described above. Future discussion of samples will refer to simple random samples unless otherwise specified. Readers should be aware that randomness of a sample is essential in order to make reasonable inferences from the results. If there is no randomness involved in drawing a sample, contact is lost with the probabilities associated with drawing an accurate inference.

SAMPLE STATISTICS

A population characteristic such as a population mean or standard deviation is called a *parameter*. A sample characteristic, on the other hand, is called a *statistic*. Statistics are typically used to make inferences about population parameters. Inferences about a population are usually inferences about its parameters. Since populations are described by their parameters, these are the values that must be inferred from the sample data. Methods for computing population parameters such as the population mean and standard deviation were covered in Chapters 3 and 4. Greek letters such as μ and σ are used to represent population parameters; Roman letters such $\bar{x}$ and s are used to represent sample statistics.

Sample Mean

The arithmetic mean of a sample of the observations of a random variable is usually designated $\bar{x}$, where x represents the random variable. Computationally, the procedure for calculating the sample mean is identical to the population mean. Formulas for calculating the sample mean for both ungrouped and grouped data are

(9.1) $$\bar{x} = \frac{\sum_{i=1}^{n} x_i}{n}$$

and

(9.2) $$\bar{x} = \frac{\sum_{i=1}^{k} x_i f_i}{n}$$

When these formulas are compared with the formulas for the population mean, the only difference occurs in the denominators. The denominator is N for the population mean, while in the formula for sample size, the denominator is n. The former represents the population size, the latter the sample size.

Sample Variance and Standard Deviation

The formulas that define the variance and standard deviation for grouped data—designated s^2 and s, respectively—appear below. Since they are identical except for the factor f_i, comparable formulas for ungrouped data are not shown.

$$(9.3) \qquad s^2 = \frac{\sum_{i=1}^{k} (x_i - \bar{x})f_i}{n - 1}$$

$$(9.4) \qquad s = \sqrt{\frac{\sum_{i=1}^{k} (x_i - \bar{x})f_i}{n - 1}}$$

Computational formulas for these sample statistics, s^2 and s, are shown below:

$$(9.5) \qquad s^2 = \frac{\sum_{i=1}^{k} x_i^2 f_i}{n - 1} - \frac{\left(\sum_{i=1}^{k} x_i f_i\right)^2}{n(n - 1)}$$

$$(9.6) \qquad s = \sqrt{\frac{\sum_{i=1}^{k} x_i^2 f_i}{n - 1} - \frac{\left(\sum_{i=1}^{k} x_i f_i\right)^2}{n(n - 1)}}$$

These equations differ from Equations 4.3 and 4.4 only in terms of the denominator. The denominator of the firm term, $n - 1$, is one less than the sample size. This value, $n - 1$, is called *degrees of freedom.*[2]

To illustrate computation of sample statistics, assume a survey of 75 national law firms has been conducted to determine average hourly rates. The hourly rates as reported on the mail questionnaire were grouped into the frequency distribution shown in Table 9.2. The mean, variance, and standard deviation were computed using Equations 9.2, 9.5, and 9.6.

$$\bar{x} = \frac{\sum xf}{n} = \frac{6{,}890}{75} = \$91.87$$

$$s^2 = \frac{\sum_{i=1}^{k} x_i^2 f_i}{n - 1} - \frac{\left(\sum_{i=1}^{k} x_i f_i\right)^2}{n(n - 1)} = \frac{730{,}700}{74} - \frac{(6{,}890)^2}{75(74)}$$

$$s^2 = 9{,}864.86 - \frac{47{,}472{,}100}{5{,}550} = 9{,}864.86 - 8{,}553.53$$

$$s^2 = 1{,}311.33$$

$$s = \sqrt{1{,}311.33} = \$36.21$$

SAMPLING DISTRIBUTIONS

The term *statistic* has been defined previously as a value computed from a sample. It is possible to draw many different samples of a given size from the same population. Therefore, any particular statistic can have different values depending upon which sample is selected. Thus, if the same statistic is computed

Table 9.2
Average Hourly Rates as Reported by 75 Law Firms, January 1983

Average Rate	Midpoint (x)	Frequency (f)	xf	x^2	x^2f
40 but under 60	50	20	1,000	2,500	50,000
60 but under 80	70	15	1,050	4,900	73,500
80 but under 100	90	10	900	8,100	81,000
100 but under 120	110	10	1,100	12,500	121,000
120 but under 140	130	8	1,040	16,900	135,200
140 but under 160	150	12	1,800	22,500	270,700
Total		75	6,890		730,700

for each of the possible samples, the value of the statistic is likely to vary from sample to sample. Theoretically it would be possible to construct a frequency distribution showing the values assumed by the statistic and their frequency of occurrence; this results in a distribution of values of a statistic. This distribution is called a *sampling distribution* because the values result from the process of sampling.

To illustrate, suppose the means of all possible samples of size n were computed from a population of size N. This would result in a set of sample means that could be arranged to form a distribution. The distribution would have its own mean, $\mu_{\bar{x}}$, and variance, $\sigma_{\bar{x}}^2$. The square root of the variance of a sampling distribution is called the *standard error* to distinguish it from the standard deviation of the population. The standard error of the mean is denoted $\sigma_{\bar{x}}$.

The same procedure could be followed in computing the various statistics from all possible samples of a given size taken from a population. For example, the variance could be computed from these various samples. Then a sampling distribution of the variance or any other statistic desired would result. Each sampling distribution has its own mean, variance, and standard error. If the characteristics of the sampling distribution of a statistic are known, the probabilities for the various possible values of the statistic can be determined.

Sampling distributions can be related to probability distributions discussed earlier. Statistics concerns itself with distributions of random variables. Probability distributions are theoretical distributions that can be used as models of populations. Probability distributions, like model airplanes or model automobiles, are not always an exact representation of the populations they represent. Nevertheless, they are convenient methods for determining probabilities of given events under certain conditions.

As noted earlier, a sampling distribution is generated by a process of sampling. Some probability distributions are approximate models for sampling distributions. The binomial distribution, discussed in Chapter 6, is a theoretical distribution that provides the probabilities of events that result from a process of sampling; in this sense it is actually a sampling distribution. The normal distribution, discussed in Chapter 7, is a theoretical distribution that can be used as a model for both population distributions and sampling distributions. For example, if one states that the means of samples are distributed normally, he implies that the normal model is also useful for describing this particular distribution that results from the process of sampling.

SAMPLING DISTRIBUTION OF SAMPLING MEANS

One of the most important sampling distributions in statistics is the distribution of the sample mean, $\bar{x}$. Assume a normally distributed population with the mean equal to μ and the standard deviation equal to σ. From this population, all possible samples of size *n* are selected, and the mean of each sample is selected. It can be shown that the set of all sample means is also normally distributed with the mean equal to μ and the standard deviation equal to σ/n.

Central Limit Theorem

The above can be extended to other distributions by use of the *central limit theorem*—a well-known mathematical theorem. This important theorem is formally described as follows:

> *If all possible random samples of size* n *are drawn from a population and the sample means computed, the mean of the sampling distribution of the means is equal to the population mean. The means of the samples drawn from a normally distributed population will themselves be normally distributed. The means of the samples drawn from all nonnormal population distributions tend to approach normality as the size of the sample is increased. The variance of the sampling distribution of the means is equal to the variance of the population divided by the size of the sample* (σ^2/n); *the standard error of the mean of the sampling distribution is, therefore,* $\sigma/\sqrt{n}$.

In practice, the distribution of $\bar{x}$ approaches a normal distribution very rapidly. Experimental evidence and advanced statistical tests have shown that, for most work, a sample size of 30 is sufficient to guarantee normality of the distribution of sample means. That is, for samples as small as 30, the assumption of the normal distribution is quite valid. When sample size is 50 or greater, the shape of the population has little practical effect on the distribution of $\bar{x}$.

To illustrate the importance of the central limit theorem in sampling, assume the mean of the population, μ, is to be estimated—the subject of Chapter 10. It is impractical (1) to take continuous samples of size *n;* (2) to compute the mean of each sample; (3) to organize the means of the samples into a sampling distribution of the means; and (4) to compute the mean of the sample means in order to estimate the population mean. The usual procedure is to take only *one* sample and, on the basis of this sample, to estimate the population mean.

How does the mean of *one* sample selected from the population compare with the population mean which is unknown? The standard deviation of the distribution of sample means, called the *standard error of the mean*. $\sigma_{\bar{x}}$, is the measure needed to answer this question.

The Standard Error of the Mean

The standard error of the mean measures *sampling errors* only. Sampling errors are errors involved in estimating a population parameter from a sample instead of including all of the essential information in the population. *Nonsampling errors* are errors resulting from bias on the part of interviewers, poorly designed questionnaires, errors in tabulation, and the like.

Assume the standard deviation of the population, σ, is known. Of course, this assumption is somewhat unrealistic because in most cases the standard deviation of the population being studied is not known. This assumption is made to present the theory underlying the formulas used to determine the reliability of sample estimates. Under this assumption, the standard error of the mean of all possible samples selected from the population, $\sigma_{\bar{x}}$, is:

$$\sigma_{\bar{x}} = \frac{\sigma}{\sqrt{n}} \tag{9.7}$$

where

σ = standard deviation of the population
n = sample size

Suppose the standard deviation of a population is known to be 80, and a sample of size 400 is drawn at random from the population. The standard error of the mean would be:

$$\sigma_{\bar{x}} = \frac{80}{\sqrt{400}} = \frac{80}{20} = 4$$

Recall that the central limit theorem states that the sample means distribute themselves normally around the population mean, μ. Thus, the following relationships hold:

1. Approximately 68 percent of all possible means of sample size 400 selected at random would be encompassed within $\mu \pm 1\sigma_{\bar{x}}$.

2. Approximately 95 percent of all possible means of sample size 400 selected at random would be encompassed within $\mu \pm 2\sigma_{\bar{x}}$.

3. Approximately 99 percent of all possible means of sample size 400 selected at random would be encompassed within $\mu \pm 3\sigma_{\bar{x}}$.

To illustrate this concept, assume a law firm is representing a client who is bringing suit against a feed manufacturer, alleging the manufacturer typically underfills sacks of feed. The law firm employs a statistician to sample the feed and determine if, in his opinion, the manufacturer is cheating on weights.

The feed manufacturer fills sacks of feed with automated machinery. The manufacturer guarantees the mean weight of the sacks to be 100 pounds with a standard deviation of 4 pounds (Thus, $\mu = 100$, $\sigma = 4$). The feed is shipped out in pallets of 64 sacks. Selecting a pallet at random, the statistician finds the average weight of the sacks to be 98.9 pounds. What percentage of the sacks should have an average weight of less than 99 pounds? Based on the central limit theorem, the means of samples of size 64 will have approximately a normal distribution; a pallet of 64 sacks can be viewed as a sample of size 64. The central limit theorem further asserts that the mean weight of all pallets of 64 sacks will be μ, 100 pounds. Finally, the central limit theorem indicates that the standard error of the mean will be

$$\sigma_{\bar{x}} = \frac{\sigma}{\sqrt{n}} = \frac{4}{\sqrt{64}} = .5$$

To determine the probability of a sample having a mean of less than 99, the statistician could use the standardized normal distribution:

$$\begin{aligned} z &= \frac{\bar{x} - \mu}{\sigma_{\bar{x}}} \\ &= \frac{99 - 100}{.5} \\ &= 2 \end{aligned}$$

From the tables for the standardized normal distribution, he would find that .4772 of the distribution would lie between the mean and two standard deviations below the mean. Therefore, there is a .4772 probability that the sample of 64 would have a mean weight of between 99 and 100 pounds and a .0228 probability that the average weight would be less than 99 pounds. The result is summarized in Figure 9.1.

In most problems involving sampling, σ is not known. When σ is not known, it can be estimated by using s, the standard deviation of the sample. One assumption made when the population parameters are not known is that sample statistics are the best estimates of their parameters. Sample statistics are, therefore, substituted for population parameters. Thus, in such cases the standard error of the mean, $s_{\bar{x}}$, is defined as

$$s_{\bar{x}} = \frac{s}{\sqrt{n}} \tag{9.8}$$

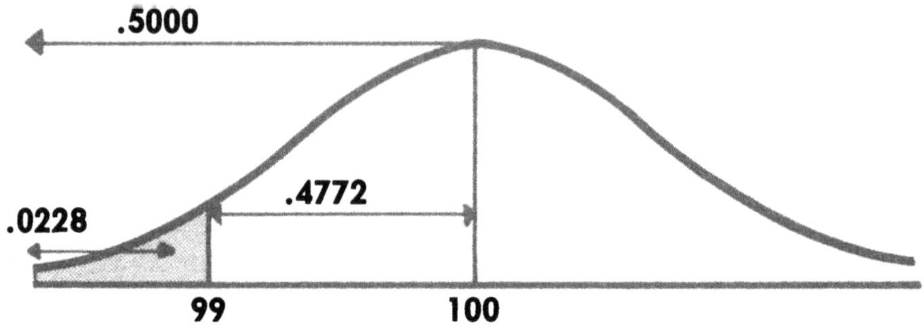

Figure 9.1
Use of the Normal Distribution to Illustrate Feed Weight Problem

where

s = standard deviation of the sample
n = sample size

The sample standard deviation, s, is determined as shown in Equations 9.4 (ungrouped data) and 9.6 (grouped data).

Finite Correction Factor for Size of the Population

The central limit theorem applies when samples are drawn from an infinite population. When samples are drawn from finite populations, a correction factor may be applied to the standard error of the mean to compensate for the size of the sample (n) relative to the size of the population (N). When sampling from finite populations, the appropriate formula for the standard error of the mean is

$$\sigma_{\bar{x}} = \frac{\sigma}{\sqrt{n}}\sqrt{\frac{N-n}{N-1}} \tag{9.9}$$

The application of the correction factor reduces the standard error of the mean. This is logical. For example, the error in estimating the mean of a population of 1,500 based on a sample of 1,000 should be less than the error based on a sample of 100 or the error based on a sample of 50. Therefore, the finite correction factor compensates for the size of the sample in relation to the size of the population. It is not usually applied unless the size of the sample is *at least* 5 percent of the size of the population.

STUDENT'S *t* DISTRIBUTION

The astute reader will have noticed by now that, in order to assume normality with respect to the distribution of sample means, the sample size must be reasonably large, i.e., at least 30. The reason for this proviso is that whenever the standard deviation of the population is *estimated from a sample* (as it usually will be), the normal distribution is only an approximation to the correct sampling distribution. The accuracy of that approximation is good only if the sample size is large.

When the standard deviation of the population is estimated from a sample—i.e., s is used as an estimate of σ—theoretically the *t distribution* should be used.[3] Also known as student's t distribution (student was the pseudonym used by W. S. Gossett, who formulated it in 1908), the t distribution is a sampling distribution. In actuality, there is not just one standardized t distribution but an entire family of them, one for each degree of freedom. In the estimation of a population standard deviation from a sample, the number of degrees of freedom is always $n - 1$, the size of the sample minus 1. The t distribution rapidly approaches the normal distribution as the number of degrees of freedom increases. When n becomes greater than 120, the t values and z values coincide. For most practical purposes, however, whenever the degrees of freedom are greater than 30, the two distributions are so similar that the normal distribution is used as an approximation to the t distribution with little significant error.

A table of the t distribution is provided in Appendix Table A.3. Note that this table provides probabilities corresponding to the areas in the right tail of the distribution. The symbol α represents the area in the tail and values corresponding to t are tabulated in the body of the table for $\alpha = .100, .050, .025, .010,$ and .005. *Degrees of freedom* appears in the extreme left column of the table; values are provided for degrees of freedom up to and including 30. The last row in the table shows the values of t when the degrees of freedom reach infinity. These values are identical to the values of z, the standardized normal distribution.

The t distribution is important in inferential procedures that involve the population mean when samples are drawn from a normal population and σ, the population standard deviation, is unknown. In those cases, σ must be estimated by s, the sample standard deviation, and the sampling distribution is treated as if it were a t distribution. Use of the t distribution instead of the normal distribution is critical when the sample size is small.

SAMPLING DISTRIBUTION OF A PROPORTION

It has been shown that the standard deviation of the distribution of means can be computed without the laborious job of taking all possible samples from the population, and the same can be done for the distribution of proportions. The formula for the true standard deviation of a proportion, called the standard error of the proportion, is presented in Equation 9.10. In conformance with the rule that parameters are designated by Greek letters, a population proportion will be represented by π and a sample proportion by p.

(9.10) $$\sigma_p = \sqrt{\frac{\pi(1 - \pi)}{n}}$$

where

σ_p = true standard error of the proportion
π = true population proportion
n = sample size

If the value of the parameter π is not known, the value from the sample may be used as an estimate, just as the value of s computed from the sample was used in making an estimate of the standard error of the mean:

(9.11) $$\hat{\sigma}_p = \sqrt{\frac{p(1 - p)}{n}}$$

where

$\hat{\sigma}_p$ = estimated standard error of the proportion
p = sample proportion
n = sample size

To illustrate, assume a survey is made in a city of 100,000 registered voters to determine voter preferences for various candidates. The survey consists of a sample of 1,500 registered voters, and it is found that 54 percent of the voters favor a particular candidate. In the absence of any other information about this proportion, the sample offers the best estimate of the parameter. The value of p from a random sample is an unbiased estimate of the value of π. Using Equation 9.11, the estimated standard error of the proportion is determined as below:

$$\hat{\sigma}_p = \sqrt{\frac{p(1 - p)}{n}} = \sqrt{\frac{.54(1 - .54)}{1,500}}$$
$$= .0001656 = .0129$$

Thus, the estimated standard error of the proportion is 1.29 percentage points. As will be noted in Chapter 10, using p as an estimate of π, it is possible to estimate the confidence interval from the information provided by the sample.

When sampling from finite populations such that the sampling fraction, n/N, is greater than 5 percent, the finite correction factor must be used for the standard error:

$$\hat{\sigma}_p = \sqrt{\frac{p(1 - p)}{n}} \times \sqrt{\frac{N - n}{N - 1}}$$

It should be pointed out that ordinarily it is more convenient to change the sample proportion to a percentage and work with this rather than to work with the proportion. This merely involves shifting the decimal point two places to the right. The arithmetic is simpler; the results are precisely the same. The following summary, therefore, is appropriate:

π = population percentage
p = sample percentage
$\hat{\sigma}_p = \sqrt{\frac{p(1-p)}{n}}$ = estimated standard error of a sample percentage

ILLUSTRATIVE USES OF SAMPLING

In *United States v. Georgia Power Company*,[4] a case previously discussed, the U.S. Circuit Court of Appeals defined the concept of a representative sample:

Where a validity study is conducted in which tests are administered to applicants, with criterion data collected later, the sample of subjects must be *representative* of the normal or typical candidate group for the job or jobs in question. This further assumes that the applicant sample is representative of the minority population available for the job or jobs in question in the local labor market.[5]

Noting the deficiencies in the testimony of the defendant's primary witness with respect to sampling, the court stated that "[i]n his testimony Dr. Hite complained of the lack of representative sample of applicants in the data with which he worked."[6]

NOTES

1. "Do Polls Tell the Story," *U.S. News and World Report*, October 5, 1964, pp. 52–53. From a copyrighted interview. Used with the permission of *U.S. News & World Report*.
2. The term $n - 1$ is used because it represents the number of *degrees of freedom* involved in estimating the standard deviation. Of the total number of values in the sample, only $n - 1$ are "free" to vary. One degree of freedom is lost when the mean is used in the calculation of the squared errors. For example, if the mean of a group of values is specified, only $n - 1$ of those values can vary without changing the mean. Assume, for instance, that the mean of five numbers is 10. Any value could be assigned to the first four numbers but there would be no freedom concerning the fifth. Its value would be fixed. Thus, one degree of freedom is lost in a sampling problem involving the standard deviation of the sample because one number (the arithmetic mean) is known.
3. Actually, the t distribution is theoretically applicable only when the *parent* population from which the sample is drawn is normal. As a practical matter, however, it works reasonably well for problems involving a variety of nonnormal parent distributions when the departure from normality is not too severe.
4. 474 F. 2d 906.
5. *Id* at 916.
6. *Id.*

SELECTED REFERENCES

Cochran, W. G. *Sampling Techniques*. 3d ed. New York: John Wiley & Sons, 1977.

Godambe, V. P. ''Foundations of Survey Sampling.'' *The American Statistician*, Vol. 24, No. 1 (February 1970), pp. 33–38.

Hansen, Morris H., William N. Harwitz and William G. Meadow. *Sample Survey Methods and Theory*. Vol. I: *Methods and Application*. Vol. II: *Theory*. New York: John Wiley & Sons, 1953.

Kish, Leslie. *Survey Sampling*. New York: John Wiley & Sons, 1965.

Solnim, Morris J. ''Sampling in a Nutshell.'' *Journal of the American Statistical Association*, Vol. 52, No. 278 (June 1957), pp. 143–61.

Yamane, Taro. *Elementary Sampling Theory*, Chap. 7. 3d ed. Englewood Cliffs, N.J.: Prentice-Hall, 1973.

CHAPTER 10

ESTIMATION

There are two major types of sampling problems involving statistical inference. One is estimating a parameter on the basis of sample evidence. This involves the estimation of μ and π. The other major type of problem involving statistical inference is testing of hypotheses. This relates to testing for the significance of difference between samples or between a sample and a population—the subject of Chapter 11. Both of these types of statistical problems revolve around the question of drawing valid conclusions concerning the population characteristic based on sample information.

The primary reasons for estimating population values from sample statistics are cost and time. The cost of a census enumeration is usually prohibitive; the time involved may be even more critical. In addition, many finite populations are too large to enumerate every time data are needed, and infinite populations preclude a complete count.

POINT AND INTERVAL ESTIMATES

Estimates can be divided into two general groups: *point estimates* and *interval estimates*. The sample mean is the "best estimate" of the parameter μ, the population mean. However, it is very unlikely that any point estimate will be exactly equal to the unknown population parameter. If additional samples of the same size are drawn from the population in exactly the same way, the value of the point estimate, $\bar{x}$, will vary from sample to sample. This variation among samples results in sampling distributions and is the cause of error in estimation. Therefore, if a single sample is taken, there is always some error present in the resulting point estimate. However, nothing in that point estimate itself indicates the magnitude of the error.

On the other hand, an interval estimate, called the *confidence interval,* includes some probability statement with respect to the degree of reliability one may place in the estimate of the parameter. That is, the amount of error in an estimate can be described in terms of an interval of values about the point estimate. The likelihood that the error will be as large as or larger than the stated interval is expressed in terms of probability. This probability is called the *confidence level* or *confidence coefficient*. For example, a confidence level of .95 theoretically means that if one were to select all possible samples of size *n* from a given population, calculate the mean for each, and construct confidence intervals for each mean, 95 percent of the intervals would include μ, the population mean. In practice, however, the term *confidence* is usually interpreted by users of interval estimates as the probability that the population mean lies in the interval.

CONFIDENCE INTERVALS FOR MEANS

It can be shown that if samples of size *n* are selected from a population and the means of all possible samples are computed, about 95 percent of the means will lie between $\mu \pm 1.96\sigma_{\bar{x}}$. This interval is called the 95 percent confidence interval. It can also be shown that if such confidence intervals were drawn for each of the sample means, 95 percent of the intervals would contain the population mean.

In practice, only one sample is usually taken. Based on the sample, an inference is drawn about a population parameter—usually the mean. To estimate the mean of the population, (1) the mean of the sample is computed; (2) a level of significance, α, is decided upon; and (3) the confidence limits are computed. The most common levels of significance are $\alpha = .10$, $\alpha = .05$, and $\alpha = .01$. The corresponding confidence levels are .90, .95, and .99, respectively. The probability statement on which the confidence interval is based is defined as

$$(10.1) \quad P(\bar{x} - ts_{\bar{x}} < \mu < \bar{x} + ts_{\bar{x}}) = 1 - \alpha$$

Or, to write it another way, the confidence limits equal $\bar{x} \pm t_{.95}s_{\bar{x}}$.

Note that to determine the confidence limits, the *t* value is used rather than the *z* value. The *t* values should theoretically be used when a sample statistic is being used to estimate a population parameter. The difference between the *z* and *t* values diminishes as the sample size increases. Selected values for both *z* and *t* at the .95 level of confidence are summarized below. Most statisticians, however, use the *z* values when the sample size exceeds 30. At this point, the estimated standard deviation of the population approximately equals the population standard deviation, and the distribution approximates a normal curve.

Sample Size	*z Value*	*t Value*
10	1.96	2.26
15	1.96	2.14
20	1.96	2.09
30	1.96	2.04
∞	1.96	1.96

Thus, if the sample size is greater than 30, the probability statement regarding the confidence interval becomes

(10.2) $P(\bar{x} - zs_{\bar{x}} < \mu < \bar{x} + zs_{\bar{x}}) = 1 - \alpha$

The sample standard error of the mean, $s_{\bar{x}}$, is used in both Equations 10.1 and 10.2 since in most problems involving sampling the standard deviation of the population is not known. In fact, the fundamental purpose of most investigations is to estimate the mean and standard deviation of the population. If, for some reason, the standard deviation of the population is known and n is large, the probability statement is

(10.3) $P(\bar{x} - z\sigma_{\bar{x}} < \mu < \bar{x} + z\sigma_{\bar{x}}) = 1 - \alpha$

Therefore, the 95 percent confidence interval estimate would be $\bar{x} \pm z\sigma_{\bar{x}}$.

To illustrate, assume a large newspaper wants to determine demographic characteristics of its readership. A sample of 400 persons is randomly selected and interviewed. One of the more important parameters to be estimated is average age of readers of the newspaper. Based on interviews with the 400 respondents, the average age is found to be 54.2 years with a sample standard deviation of 12.8 years. Assume the sample standard deviation has been corrected for loss of one degree of freedom and therefore represents an estimate of the population standard deviation. The researchers wish to construct a 95 percent confidence interval for the mean age of readers. The standard error of the mean, $s_{\bar{x}}$, is calculated

$$s_{\bar{x}} = \frac{s}{\sqrt{n}} = \frac{12.8}{\sqrt{400}} = .64$$

Now the confidence interval (Equation 10.2) is

$$P(\bar{x} - zs_{\bar{x}} < \mu < \bar{x} + zs_{\bar{x}}) = 1 - \alpha$$

Or the confidence interval may be viewed as $\bar{x} \pm zs_{\bar{x}}$.

$$\begin{aligned} \text{Confidence interval} &= \bar{x} \pm zs_{\bar{x}} \\ &= 54.2 \pm 1.96(.64) \\ &= 54.2 \pm 1.3 \\ &= 52.9 \text{ and } 55.5 \end{aligned}$$

Thus, the .95 confidence limits are 52.9 years and 55.5 years, and .95 confidence interval is the distance between these two values.

The statistician can now make a statement that at the .05 level of significance, the mean age of the population of readers is somewhere in the interval between 52.9 years and 55.5 years. To put it another way, the statistician is 95 percent confident that the population mean is in the interval between 52.9 and 55.5 years. He is not sure, however, that the population mean is within that interval.

The sample mean from the survey (54.2 years) could be one of the sample means which includes the population mean in the 95 percent confidence interval, or the sample mean being computed could be one of the 5 of 100 which does not include the population mean within the confidence interval. This, of course, is one of the risks involved in sampling.

CONFIDENCE INTERVALS FOR PROPORTIONS

Another important facet of sampling is making inferences about the population proportion based on the sample proportion. In the survey noted earlier regarding readership, one of the most important issues to be resolved is extent of readership of the paper. Of the 400 persons interviewed, 220 indicated they read the paper on a regular basis. The publisher, however, wants to know between what limits one would expect the true proportion of readers to lie.

The statistician in this case used the .95 confidence level. Since π is not known, the formula for the estimated standard error of the proportion (Equation 9.11) must be used:

$$\hat{\sigma}_{\bar{p}} = \sqrt{\frac{p(1-p)}{n}}$$

In this example,

$$\hat{\sigma}_{\bar{p}} = \sqrt{\frac{(.55)(.45)}{400}} = .025$$

The 95 percent confidence interval estimate is equal to $.55 \pm 1.96(.025) = .500$ to .599. Thus, the interval estimate of readership would range from 50.0 percent to 59.9 percent or ± 4.9 percentage points. The point estimate would be 55 percent.

Interval estimates for proportions are analogous to interval estimates for means. If many samples of size 400 were taken from this population, their sample proportions computed, and intervals constructed, 95 percent of these intervals would include the true proportion, and 5 percent would not.

SAMPLE SIZE

One of the most common questions asked about the estimation process concerns sample size. How many items should be sampled from the population? In examining this question, several observations are in order. In estimating a population proportion or population mean, it is obvious that one can improve the precision of the estimate or size of the error—i.e., the width of the confidence interval—by giving up some probability of being right. For example, the width of the interval may be too great; this can be reduced only by lowering the confidence level. One can improve the reliability—the probability of being right—on the other hand, by giving up some precision, i.e., by increasing the interval in which it is inferred that the true population value lies. That is, to increase reliability from the .90 confidence level to the .95 would increase the confidence interval, thus reducing the precision. The only way to increase both precision and reliability is to increase sample size.

The points discussed can be summarized in the following relationship:

$$E = K(SE)$$

where

$$E = \text{error}$$
$$K = \text{confidence level}$$
$$SE = \text{standard error}$$

That both precision and reliability can be increased by increasing the sample size (n) is apparent from the fact that n appears as the denominator in both the standard error of the mean and the standard error of the proportion. This means that as sample size increases, the standard error decreases. For a given sample size, the desire for reliability and the desire for precision are tradeoffs; the manner in which the results are stated always involves a compromise between precision and reliability. To satisfy both demands more fully, the sample size must be increased, thus involving additional costs. Decision makers must continuously weigh additional costs of larger samples against the increased precision and reliability.

It should be noted that doubling a sample size *does not* halve the error of the estimate. Remember the standard error value is proportional to the reciprocal of the *square root* of the sample size $\sqrt{n}$, not n. Thus, to double the precision of the estimate of the parameter—i.e., reduce its standard error by a factor of 2—sample size must be increased not by a factor of 2 but by its square, 4. This illustrates the *principle of decreasing variation:* the size of the sampling error decreases as the reciprocal of the square root of n, the number of items in the sample, increases.

Another question often asked regarding sample size relates to the magnitude of the population from which the sample is drawn. Should the size of the population be known? Contrary to what logic would dictate, the size of the population has no effect on the size of the sample as long as $N > 20n$. This point has been repeatedly made by national pollsters, who point out that no larger sample is needed for polling the United States than, say, New York City.

Sample Size for Means

In estimating the mean of a population, the error in the estimate is $E = |\bar{x} - \mu|$ or the absolute difference in the mean of the sample and population. Since $z = (\bar{x} - \mu)/\sigma_{\bar{x}}$ and since $\sigma_{\bar{x}} = \sigma/\sqrt{n}$, E can be defined as

$$E = \frac{z^2\sigma^2}{n}$$

This equation can be solved for n, resulting in a sample size equation.

$$(10.4) \qquad n = \frac{z^2\sigma^2}{E^2}$$

E is specified as the maximum error that can be tolerated with the selected confidence and is determined in advance. z is determined by the level of confidence desired. Thus, n and σ are the only unknowns in the sample size equation. It is, however, impossible to solve a single equation with two unknowns. Therefore, to solve for a value of n, σ must be either known or estimated.

It is not likely that the standard deviation of a population will be known if the mean is unknown, but there are occasions when the standard deviation might be assumed to be known on the basis of other data such as previous surveys. If there are no data available and σ is not known, frequently a small pilot sample is taken for the purpose of obtaining an estimate of σ. This estimate is used in the sample size equation for estimating n.

To illustrate the computational procedures involved in estimating sample size, assume a large metropolitan bar association is interested in surveying its membership to determine mean earnings with 95 percent confidence. The bar association has specified that the error in the estimate must not exceed \$500. From previous surveys of mean earnings conducted by other bar associations, the statistician responsible for the study knows that the standard deviation of incomes is generally about \$5,000. The required sample size (Equation 10.4) is estimated as follows:

$$n = \frac{(1.96)^2(5{,}000)^2}{(500)^2} = \frac{96{,}040{,}000}{250{,}000} = 384.16$$

The minimum required sample size is 384.16 or, rounding to the next higher whole unit, 385.

Sample Size for Proportions

The procedure for estimating the appropriate sample size is similar to the one used for means. The error in the estimate is defined as $E = |p - \pi|$. Since $z = (p - \pi)/\sigma_p$ and since $\sigma_p = \sqrt{\pi(1 - \pi)/n}$, E can be defined as:

$$E = z\sqrt{\frac{\pi(1 - \pi)}{n}}$$

Solving for n results in the sample size equation for proportions:

$$(10.5) \qquad n = \frac{z^2\pi(1 - \pi)}{E^2}$$

Thus, n or sample size depends upon the value of p, the parameter which is to be estimated through the sampling procedure. Many times the researcher has no idea what the proportion might be. In order to be conservative and not underestimate n, the simplest procedure is to substitute that value of π which will provide the maximum value of n. When $p = .50$, the product $\pi(1 - \pi)$ is a maximum. Substituting .5 for π in the sample size equation results in the largest possible value of n. Making this substitution reduces Equation 10.5 to

$$(10.6) \qquad n = \frac{.25z^2}{E^2}$$

To illustrate the sample size equation for proportions, recall the previous example for the newspaper publisher who wished to determine readership patterns of his paper. In that example, 400 persons were interviewed; the sampling error for readership was found to be ±4.9 percentage points at the 95 percent confidence level. Assume the publisher wants to reduce the sampling error to ±3.0 percentage points. How large a sample should be taken? Using a value for E of .030 and substituting in Equation 10.6, the required sample size is

$$n = \frac{.25(1.96)^2}{(.030)^2} = 1{,}067.1 \text{ or } 1{,}068$$

To reduce sampling error from ±4.9 percentage points to ±3.0 percentage points would require a sample size of 1,068.

ILLUSTRATIVE USES OF ESTIMATION

Many of the case histories dealing with estimation relate to sample size, particularly the adequacy or inadequacy of a given sample size. The courts have been mixed in applying standards to sample size. In many instances, a given sample size was considered invalid while a smaller sample was held to be valid.

In *Keely v. Westinghouse Electric Corp*,[1] a case heard by the U.S. district court, the court held that:

> while statistics may be used to establish a prima facie case . . . the statistics produced herein fail to establish the same. Defendant's plant is small, employing only 34 persons. . . . This amount does not justify the conclusion that defendant violated Title VII. . . . Without further evidence, plaintiff's statistics concerning departments in a company employing only 34 persons are insufficient to establish a violation of Title VII. . . . In *Harper v. Trans World Airlines, Inc.* . . . the court held that "statistical evidence derived from an extremely small universe . . . has little predictive value and must be discarded."[2]

Ochoa v. Monsanto Company[3] represents another example of a court mentioning sample size in its ruling. Hearing this case on appeal from the district court, which ruled in favor of the defendant, the U.S. court of appeals held that, in light of the smallness of the numbers and absence of any specific probative evidence, the trial judge did not err in finding no discrimination:

[W]e conclude in this instance that the extremely small universe involved and the indication from the existing statistical information did not preclude the trial court from finding an absence of discrimination . . . the smallness of the number demonstrates that the trial court was not compelled to allow such statistical showing to set in train the usual presumptions or make a finding of preference thereon.[4]

A third case where a court ruled that sample size was too small is *Mayor of Philadelphia v. Educational Equality League*,[5] a case heard by the Supreme Court on appeal from the U.S. court of appeals. The Supreme Court, overturning the verdict of the court of appeals in favor of the defendant, held:

[T]he simplistic percentage comparisons undertaken by the Court of Appeals lack real meaning in the context of this case. . . . Furthermore, the District Court's concern for the smallness of the sample presented by the 13-member panel was also well founded. The Court of Appeals erred in failing to recognize the importance of this flaw in straight percentage comparisons. . . . In our view, this type of proof is too fragmentary and speculative to support a serious charge in a judicial proceeding.[6]

On the other hand, small samples have been readily accepted by some courts. In *Chicano Police Officers' Ass'n v. Stover*,[7] the U.S. court of appeals vacated and remanded the decision of the district court, holding that the fact that only 26 Chicano examinees took a particular test did not make the results of the test irrelevant on the theory that the sample was so small as to be statistically insignificant. The court pointed out:

We must disagree with the ultimate findings and conclusions of the trial court. The smallness of the sample should not be grounds here for rejecting the proof. If it were, the tendency would be to deny employees in small plants the type of protection the civil rights statutes afford. . . . And, in any event, we feel the group tested was not too small to be evaluated as significant.[8]

NOTES

1. 404 F. Supp. 573 (1975).
2. *Id* at 578–79.
3. 473 F. 2d 318 (1973).
4. *Id* at 319–20.
5. 415 U.S. 605 (1974).
6. *Id* at 620–21.
7. 526 F. 2d 431 (1975).
8. *Id* at 439.

SELECTED REFERENCES

Fisher, R. A. *Statistical Methods and Scientific Inference*. London: Oliver & Boyd, 1956.

Guenther, William C. *Concepts of Statistical Inference*, Chap. 5. 2d ed. New York: McGraw-Hill Book Co., 1973.

______. "Shortest Confidence Intervals." *The American Statistician*, Vol. 23 (February 1969), pp. 22–25.

Leabo, Dick A. *Basic Statistics*, Chap. 9. 4th ed. Homewood, Ill.: Richard D. Irwin, 1972.

Miller, Irwin and John E. Freund. *Probability and Statistics for Engineers*, Chap. 8. Englewood Cliffs, N.J.: Prentice-Hall, 1965.

Savage, Leonard J. *The Foundations of Statistics*, Chaps. 15, 17. New York: John Wiley & Sons, 1954.

CHAPTER 11

INFERENCE: TESTS OF SIGNIFICANCE

Up to this point, the discussion of sampling has involved estimation of certain characteristics of the population from which the sample was drawn. It was pointed out that estimates of values for the population are subject to error. Nevertheless, limits may be determined within which, with a given degree of confidence, the population values may be expected to fall.

Estimates from samples are only approximate; this fact must be taken into consideration in making decisions based on such estimates. If, however, population data are used as the basis for comparison, the data can be considered accurate. Recall the initial example in Chapter 4. The same six attorneys worked on two cases. Case A required a total of 100 billable hours; case B involved 110 hours. The difference in hours worked is small. If the records were checked sufficiently to be certain they were accurate, there is no question but there was a difference between hours worked on the two cases.

If averages, for example, are based on samples, they are approximate, and it would not be immediately clear whether the difference between two averages was a real difference or was due to the approximate nature of the results. In such a situation, the analysis of sample data introduces an element not present in population data. When data based on a complete enumeration are used, there is always the possibility that different kinds of mistakes have been made in compiling the data. If careful work has been done, there is usually no reason why such mistakes cannot be eliminated. This is not true with respect to sampling error. It is inevitable that variation will exist between the results of a complete enumeration and a sample even though care has been taken to ensure accuracy in the collection and analysis of the data. Since a properly designed sample provides the confidence interval of the estimate, there is enough information to permit drawing conclusions even though the information is only approximate.

RATIONALE FOR HYPOTHESIS TESTING

A hypothesis is an assumption to be tested. The statistical testing of hypotheses is probably the most important technique in statistical inference. The basis for hypothesis testing, frequently referred to as statistical decision making, is noted below.

If a random sample is drawn from a population, one could reasonably assume that the sample will resemble the population. Estimates of unknown parameters through use of sample statistics are made on the basis of this assumption. If an assumption is made about the value of a population parameter, it is reasonable to expect that an appropriate statistic computed from that sample will have a value that is reasonably close to the hypothesized value. Such is true, however, only if the hypothesized value is correct and if the statistic has the properties that make it a good estimator of the pertinent parameter. A *test statistic* is used to test a hypothesis.

As noted earlier, statistics are random variables; they have sampling distributions that exhibit the property of variation. Thus, even if the hypothesized value is correct, it would not be expected that the statistic would have exactly the same value as the parameter. Some difference is likely due to chance. For example, if there is a "relatively small difference" between the mean (or proportion) of a sample and the established mean (or proportion), it is clear from knowledge of sampling variation that the observed difference easily may have occurred by mere chance. If the value of the statistic deviates greatly from the hypothesized parameter value, there is reason to doubt that the hypothesis is correct; it is probable that the sample did not come from the population with the assumed characteristic. The greater the difference between the value of the statistic and the hypothesized value, the greater the doubt about the correctness of the hypothesis.

In statistics, it is necessary to specify, in quantitative terms, exactly what constitutes a significant deviation. In testing, significance is specified in terms of the probability that a particular deviation occurs by chance given the hypothesized value of the parameter is correct. If the properties of the sampling distribution of the test statistic are known, the probability of a deviation of any magnitude is easily determined.

PROCEDURE FOR HYPOTHESIS TESTING

A test of hypothesis, like other statistical techniques, is not guaranteed to yield the correct answer 100 percent of the time. In an uncertain world there is no chance of having such occur. If one recognizes that the world is uncertain, the next best thing is to control or, as a minimum, know the probability of making an error. A statistical test of hypotheses will accomplish this, assuming the statistician follows the correct procedure. The general procedure for hypothesis testing is summarized below.

1. Establish the null and alternate hypotheses.
2. Select a significance level for the test.
3. Specify the test statistic.

4. Formulate a decision rule.

5. Draw a random sample of size n from the population and compute the value of the test statistic.

6. Accept or reject the null hypothesis according to the decision rule.

The Null Hypothesis

The hypothesis being tested is called the *null hypothesis,* designated H_0. Since statistical hypotheses are usually assumptions about the value of some unknown parameter, the hypothesis specifies a numerical range or ranges of values for the parameter. These stated values, which are accepted or rejected on the basis of the test, constitute the null hypothesis. If the condition specified under the null hypothesis is rejected by the test, the condition is assumed to be false. It therefore follows that some other condition is assumed to be true. Rejection of the null hypothesis is equivalent to acceptance of an *alternate* hypothesis, designated H_1. There are always two hypotheses in any hypothesis test—the null hypothesis and the alternate hypothesis.

The null hypothesis can be expressed in either of two general forms. In testing a hypothesis about the value of an unknown population mean, for example, the null hypothesis may state that the mean is equal to some specific value. The alternate hypothesis states that the population mean is not equal to the value stated in the null hypothesis. Thus,

$$H_0: \quad \mu = 15$$
$$H_1: \quad \mu \neq 15$$

A second general form for testing hypotheses is to establish a null hypothesis stating (1) that the unknown parameter is equal to or greater than some value, or (2) that it is equal to or less than the specific value. For example,

$$H_0: \quad \mu \geq 15$$
$$H_1: \quad \mu < 15$$

or

$$H_0: \quad \mu \leq 15$$
$$H_1: \quad \mu > 15$$

It should be noted that common practice is to establish the null hypothesis so that acceptance requires no positive action, while rejection leads to some positive act.

Level of Significance, α

In practice, three of the most commonly used levels of significance are the .10 level, the .05 level, and the .01 level. These respond to the confidence coefficients of .90, .95, and .99, respectively. The choice as to level of significance is largely a matter of judgment.

In selecting α, the statistician must decide on the probability that any difference between a sample and a hypothesized population, for example, could have occurred by chance alone. That is, he must decide on precisely how small the chance must be before the difference is arbitrarily called significant and the null hypothesis is rejected. When the hypothesis is rejected, the statistician concludes that the sample came from a population that has a parameter different from the hypothesized value. For example, if he determines that the probability is 6 out of 100 (.06), the statistician may not reject the hypothesis, i.e., he accepts it. But if the probability of a chance occurrence is 5 out of 100 (.05) or less, he may begin to believe that the chance factor is probably not strong enough to explain the observed difference. Some factor other than chance has been operating to cause the difference. In this case, the statistician concludes that the difference between the sample result and the hypothesized population is significant, leading to a rejection of the null hypothesis and acceptance of the alternate hypothesis.

Test Statistic

Once the level of significance has been decided upon, the test statistic must be selected. In most instances, hypotheses are tested by defining the decision rule in relation to either the normal or student's t distribution, depending upon sample size. Critical values of z and t can be used for the test. After the sample is taken, the sample statistic is converted to a corresponding value of z or t which is compared to the critical value of z or t to make the decision.

Decision Rule

The next step in the hypothesis testing procedure is to formulate a decision rule. A decision rule is a statement that indicates the action to be taken—e.g., accept H_0 or reject H_0—for all possible values of the test statistic. The decision rule divides the area under the curve of the sampling distribution into two mutually exclusive regions. These regions are called the *acceptance region* and the *rejection region,* and are illustrated in Figure 11.1. If the sample result is within the acceptance region, the null hypothesis is accepted, and any difference between the sample value and the hypothesized value is assumed to be due to chance. If, on the other hand, the sample result is outside the acceptance region, the null hypothesis is rejected and the alternate hypothesis is accepted. The difference between the two values is said to be significant at the specified level, α.

As was discussed in Chapter 9, 99 percent of the area under the normal curve falls within ± 2.58 standard deviations of the mean. If the computed statistic z falls within $\pm 2.58\ \sigma$, it is assumed that any difference between the sample value and the population value was due to chance. If the computed z value is 2.58 or

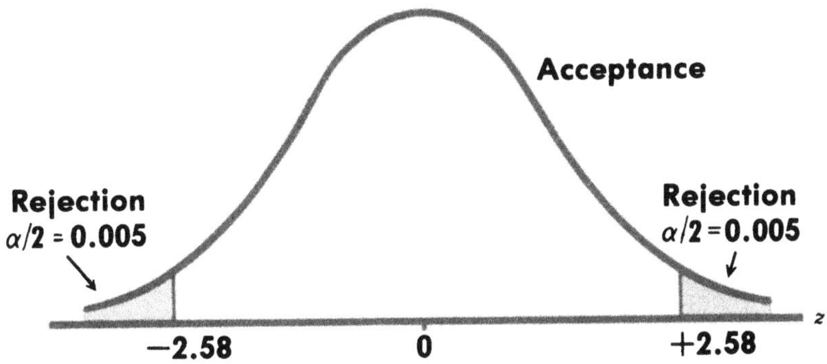

Figure 11.1
Two-Tailed Test, Areas of Acceptance and Rejection, $\alpha = .01$

greater, the null hypothesis is rejected. It is rejected on the basis that it is highly unlikely that a computed z value would be 2.58 or greater by chance.

It was previously pointed out that there are two general forms of null and alternate hypotheses in statistical tests. The first form tests the null hypothesis that a parameter, such as μ, has a value that is exactly equal to some specific value against the alternative hypothesis that it is not equal to that value, e.g.,

$$H_0: \quad \mu = 15$$
$$H_1: \quad \mu \neq 15$$

If the value of the test statistic deviates significantly from μ in either direction, H_0 is rejected. Therefore, it is necessary to provide for a rejection region in both tails of the sampling distribution of the test statistic. This type of test, illustrated in Figure 11.1, is called a *two-tailed test.* The rejection regions are selected so that an area corresponding to $\alpha/2$, where α is the significance level, is placed in each end of the sampling distribution. Half of the rejection area lies in each tail.

When only one tail of the sampling distribution is used, the test is described as a *one-tailed test* (Figure 11.2). This test corresponds to the second general form of the null and alternate hypotheses. This form tests the hypothesis that the parameter value μ is equal to or greater than some value, or that it is equal to or less than the value.

$$H_0: \quad \mu \geq 15$$
$$H_1: \quad \mu < 15$$

or

$$H_0: \quad \mu \leq 15$$
$$H_1: \quad \mu > 15$$

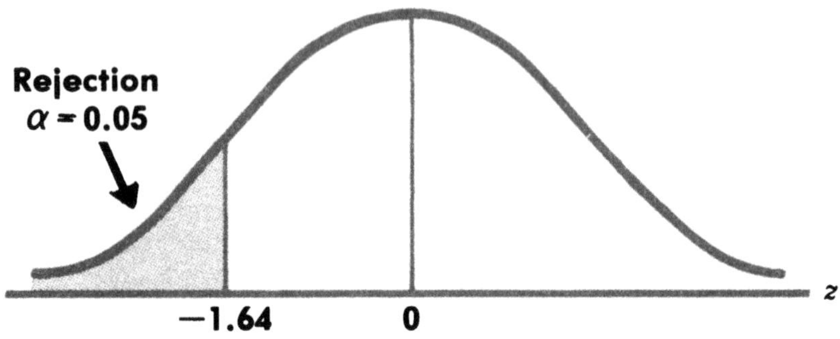

Figure 11.2
Rejection Area for One-Tailed Test, $\alpha = .05$

These hypotheses result in a one-tailed test since the null hypothesis is rejected for a significant deviation in one direction only. Therefore, the entire rejection region corresponding to α is located only in one tail of the sampling distribution of the statistic; the tail containing the rejection region is always in the direction of the inequality that appears in the alternative hypothesis. For example, if the hypothesis is tested that μ is equal to or greater than 15, the alternate hypothesis is that μ is less than 15. The rejection region is in the left tail. When $\alpha = .05$, the hypothesis will be rejected if the calculated value of z is less than or equal to -1.645 and accepted if z is greater than -1.645.

Sample and Sample Results

Random samples are selected as was discussed in Chapter 9; sample size is determined as shown in Chapter 10. Various statistics, such as the mean and standard deviation, are calculated as illustrated in earlier chapters.

The Decision

The decision to accept or reject a null hypothesis depends upon the decision rule. Once the decision rule has been established, it is a simple matter of determining whether the calculated value of the sample statistic falls into the acceptance or rejection region of the curve of the sampling distribution.

ERRORS IN HYPOTHESIS TESTING

In testing hypotheses, it would be desirable if the test always resulted in the acceptance of the null hypothesis when it is true and rejection when it is not true. As long as the decision to accept or reject a hypothesis is based on sample data, there is no testing procedure that will always lead to the correct decision. An incorrect decision is an error. There are two possible types of errors associated with a wrong decision; these are labeled *type I error* and *type II error*. Both kinds of error are illustrated in Table 11.1.

Table 11.1
Errors in Hypothesis Testing

	Decision	
Condition	*Accept* H_0	*Reject* H_0
H_0 is true	no error	type I error
H_0 is false	type II error	no error

A type I error, represented by α, is rejecting a null hypothesis when in fact it is true. The probability of a type I error is equal to the alpha risk, α, and is therefore controlled. A type II error, represented by β, is accepting a null hypothesis when in fact it is false. The probability of committing a type I error can be reduced by the selection of a strict level of significance, e.g., an α of .01 or .001. Thus, in selecting a significance level for a test of a hypothesis, the risk the statistician is willing to assume of rejecting a true null hypothesis is specified by α.

It should be pointed out that as the risk of a type I error is decreased, the risk of a type II error is increased commensurately. There are, in other words, tradeoffs involved. The acceptance interval is broad for a strict level of significance and may include differences caused by other than random factors. This increases the probability of accepting the null hypothesis when it is false. On the other hand, if a less demanding level of significance is chosen (e.g., α of .05 instead of, say, .01), the probability of making a type II error decreases, but the probability of making a type I error increases. The only way to reduce the probability of both kinds of errors is to increase the sample size. Neither, of course, can ever be completely eliminated.

An excellent example of type I and type II errors is found in our legal system. In our criminal justice system, type I and type II errors come into play every time a defendant goes on trial; a null hypothesis is tested against an appropriate alternate hypothesis. Since the null hypothesis always presumes that the defendant is innocent, the criminal court procedure may be stated as:

H_0: Defendant is innocent
H_1: Defendant is guilty

Since type I errors occur when a true hypothesis is rejected, finding a defendant guilty when he is in fact innocent is an error of this type. Finding a defendant innocent when he is guilty is a type II error. The reader can judge as to which type of error is the more serious.

TESTING DIFFERENCES BETWEEN MEANS

Through hypothesis testing, observed differences between a sample mean (proportion) and a hypothesized population mean (proportion) or between two sample means (proportions) are studied to evaluate the probability that they occurred by chance alone. Several types of test situations, involving both means and proportions, are discussed in this section.

Population Mean and Sample Mean, σ Known

This represents a typical problem in quality control. Assume, for example, a law firm is retained by a trucking firm in a case involving warranties on new truck tires. Experience over the years by the truck tire manufacturer has revealed that, for a certain model of tire, the mean life is 40,000 miles with a standard deviation of 1,600 miles—i.e., $\mu = 40{,}000$, $\sigma = 1{,}600$. The trucking firm claims that the last 100 tires it purchased from the manufacturer averaged 39,000 miles—i.e., $\bar{x} = 39{,}000$—and that the manufacturer is guilty of false advertising when he states the average life of new tires is 40,000 miles. To develop its case, the law firm representing the trucker employs a statistician to serve as an expert in this area. In developing the case, the statistician followed essentially the steps outlined earlier in this chapter.

The null and alternate hypotheses, assuming a significance level (α) of .01, are as follows:

H_0: $\mu = 40{,}000$	The mean of the population from which the sample came equals the hypothesized parameter, 40,000 miles. The observed difference, 1,000 miles, is not significant; it occurred by chance alone.
H_1: $\mu \neq 40{,}000$	The mean of the population from which the sample came does not equal the hypothesized parameter. The observed difference is significant; it is too great to have occurred by chance alone.

In the above case, a two-tailed test is implied since the hypothesis is tested in both directions. Since the distribution of the sample is approximately normal, .01 of the means fall in the two tails of the distribution delineated by $\pm 2.58\sigma$ as shown in Figure 11.1. Thus, the probability of committing a type I error, rejecting a true hypothesis, is .01.

Because the distribution is assumed to be normal, the sampling distribution of the differences between $\bar{x}$ and μ is normally distributed with a mean of 0 and a standard error of the mean equal to $\sigma/\sqrt{n}$. The statistic z is needed. The calculated value of z is

$$z = \frac{\bar{x} - \mu}{\sigma_{\bar{x}}}$$

That is, after the sample is taken, the sample statistic is converted to a corresponding value of z which is compared to the critical or tabular value of z to make a decision.

In this example, the test statistic is $\bar{x}$ or 39,000. The standard error of the mean, $\sigma_{\bar{x}}$, is equal to 160 (1,600 divided by 10, the square root of n). The decision rule is defined with respect to the distribution of

$$z = \frac{\bar{x} - \mu}{\sigma_{\bar{x}}}$$

which has a normal distribution. The critical or tabular value of z (Table A.2) is ± 2.58. The decision rule states:

> *A random sample of* n = *100 is drawn from the population. The sample mean,* x̄, *is converted to a value of* z. *If the calculated value of* z *is greater than or equal to 2.58 or less than or equal to −2.58,* H_0 *is rejected. Otherwise* H_0 *is accepted.*

Thus, z in the tire example is defined as

$$z = \frac{39{,}000 - 40{,}000}{160} = -6.25$$

Since the computed z value of 6.25 is greater than the critical value of 2.58, H_0 is rejected. H_1 is therefore accepted. There is a significant difference between the sample mean of 39,000 miles and the manufacturer's advertised or hypothesized mean of 40,000 miles. It is concluded that the sample did not come from a population with $\mu = 40{,}000$. That is, the observed difference is due to other than chance occurrence. This test is illustrated in Figure 11.3.

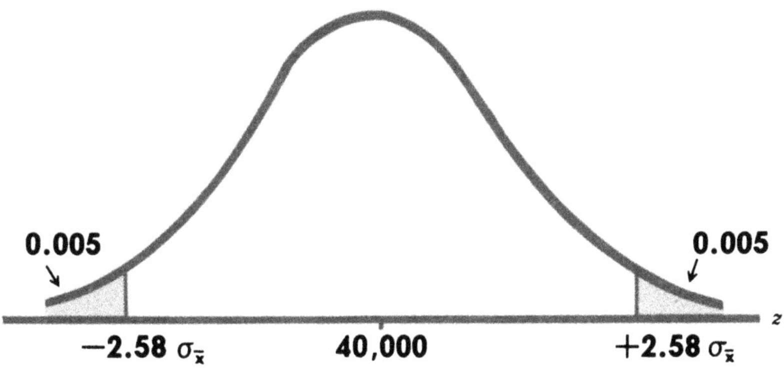

Figure 11.3
Test of Difference between Population Mean and Sample Mean

Alternatively, the statistician could have used a one-tailed test to determine if the difference between the sample and population mean was significant. The relevant hypotheses are as follows:

H_0: $\mu \geq 40{,}000$	The mean of the population from which the sample came is equal to or greater than the hypothesized parameter, 40,000 miles. The observed difference, 1,000 miles, is not significant; it occurred by chance alone.
H_1: $\mu < 40{,}000$	The mean of the population from which the sample came is less than the hypothesized parameter, 40,000 miles. The observed difference is significant; it is too great to have occurred by chance alone.

Using the 1 percent level of significance, the critical value of z is 2.33. The computed value of z, which has been found to be 6.25, falls into the rejection area. The null hypothesis is rejected and the alternate is accepted. It is therefore concluded that the sample of tires came from a population with a mean of less than 40,000 miles. This is illustrated in Figure 11.4.

Population Mean and Sample Mean, σ Unknown

In the previous example, the standard deviation of the population was known. There are many situations, however, in which such is not the case. In Chapter 10, the standard deviation of the sample was used to estimate the population standard deviation in computing the confidence interval of an estimate. The same approach can be used in testing of hypotheses when the population standard deviation is not known.

Assume now that a law firm is representing a trucker who purchased 100 tires from a tire manufacturer who has advertised in various media that the tires have an average life of 40,000 miles. The trucker has kept careful records on each tire; the mean life of the 100 tires was 39,000 miles. Using these records, the statistician retained by the law firm representing the trucker determines that the standard deviation of the sample of 100 tires was 4,200 miles. From this sample data,

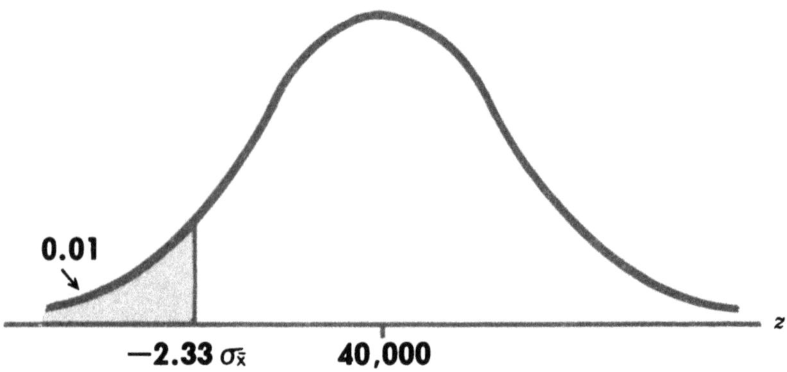

Figure 11.4
Test of Difference between Population Mean and Sample Mean

the statistician wants to determine if the mean life of the 100 tires differs significantly from 40,000 miles, the hypothesized population mean. The only information known about variation in life of tires is the standard deviation of the sample; the standard deviation of the population is unknown.

As noted earlier, the standard deviation of the sample can be used to estimate the standard deviation of the population, i.e., $s_{\bar{x}} = s/\sqrt{n}$. It must be remembered, however, that the estimate of the standard deviation is only approximate since it contains sampling error the same as any other estimate of a parameter from a statistic. The estimated standard error of the mean, based on the sample of 100 truck tires, is 420 miles. The problem is summarized as follows:

H_0: $\mu \geq 40{,}000$, H_1: $\mu < 40{,}000$; $\bar{x} = 39{,}000$, $s = 4{,}200$, $s_{\bar{x}} = 420$. With $\alpha = .01$, the critical value of z is -2.33. Thus,

$$z = \frac{\bar{x} - \mu}{s_{\bar{x}}} = \frac{39{,}000 - 40{,}000}{420} = -2.38$$

Since the calculated value of z is less than the tabular value, H_0 is rejected and H_1 is accepted. It is concluded, at the .01 level, that the mean life of tires produced by the manufacturer is less than 40,000 miles—the advertised life.

To illustrate testing the difference between a population and sample mean when the sample size is small, suppose several buyers of wire cable are threatening to bring suit against the manufacturer of the cable. The buyers charge that the product, sold and advertised as having a breaking strength of 200 pounds per square inch (psi), does not meet breaking strength standards. Attorneys representing the defendant retain a statistician to perform independent tests on the breaking strength of the cable; the purpose of the tests is to determine whether mean breaking strength differs significantly from the advertised strength of 200 psi.

A random sample of 25 sections of the cable shows a mean breaking strength of 198 psi and a standard deviation of 12 psi. The statistician must determine whether the difference in mean breaking strength occurred by mere chance or is, in fact, caused by factors other than chance. In the case of the former, the difference is not significant; in the latter case, it is significant. The hypotheses to be tested at the .05 level of significance are

$$H_0: \quad \mu \geq 200$$
$$H_1: \quad \mu < 200$$

Since the sample size is less than 30 and σ is unknown, the test statistic is t:

(11.1) $$t = \frac{\bar{x} - \mu}{s_{\bar{x}}}$$

Remember that the sample standard error of the mean (Equation 9.8) is

$$s_{\bar{x}} = \frac{s}{\sqrt{n}} = \frac{12}{5} = 2.4$$

The tabular value of t at $n - 1$ or 24 degrees of freedom is 2.064. Since this value is less than the critical or calculated value of t, H_0 is accepted. It is therefore concluded that the difference between the sample mean breaking strength of 198 psi and the hypothesized mean of 200 is due to chance alone.

Two Sample Means—Large Samples

Particularly in experimental work and in quality control, the means of two samples are tested to determine if there is a significant difference between them. Such problems merely involve the sampling distribution of the *differences* between means. If independent random samples are taken from each population and if the means of each sample are computed, then $\bar{x}_1$ is the mean of the first sample and $\bar{x}_2$ is the mean of the second. Since the two samples should be representative of their respective populations, $\bar{x}_1$ would be approximately equal to μ_1, and $\bar{x}_2$ approximately equal to μ_2. Further, $\bar{x}_1 - \bar{x}_2$ would be approximately equal to $\mu_1 - \mu_2$. Defining $d^* = \bar{x}_1 - \bar{x}_2$ and if $\mu_1 = \mu_2$, the expected value of d^* is zero. That is, $E(d^*) = E(\bar{x}_1 - \bar{x}_2) = \mu_1 - \mu_2 = 0$.

If $\bar{x}_1$ and $\bar{x}_2$ have normal distributions, either because the respective populations are normal or because of the central limit theorem, then the distribution of d^* is also normal, with a mean equal to $\mu_1 - \mu_2$. The variance of d^* is

$$(11.2) \qquad \sigma^2_{d^*} = \frac{\sigma_1^2}{n_1} + \frac{\sigma_2^2}{n_2}$$

where

$\sigma_1{}^2$ = variance of first population
$\sigma_2{}^2$ = variance of second population
n_1 = size of first sample
n_2 = size of second sample

The standard deviation of d^*, the square root of the variance, is defined as

$$(11.3) \qquad \sigma_{d^*} = \sqrt{\frac{\sigma_1^2}{n_1} + \frac{\sigma_2^2}{n_2}}$$

The hypotheses for the test of difference between two means are

H_0: $\mu_1 = \mu_2$ or $\mu_1 - \mu_2 = 0$
H_1: $\mu_1 \neq \mu_2$ or $\mu_1 - \mu_2 \neq 0$

After a random sample is drawn from each population, the mean of each sample is calculated and the value of the statistic $d^* = \bar{x}_1 - \bar{x}_2$ is determined. $E(d^*) = 0$ if the null hypothesis is true. If d^* deviates significantly from 0 in either direction, H_0 should be rejected. As was the case in testing the difference between sample and population means, if the critical value, d_c^*, exceeds the calculated value, d^*, then the null hypothesis should be rejected and the alternate accepted. The two-tailed test is always used in testing differences between means. In order to establish a decision rule, α, the level of significance, must be specified.

In testing differences between sample means, σ_1 and σ_2 are almost always unknown. Common practice, therefore, is to use the standard deviations of the sample means to estimate the population standard deviations. Thus, the standard error of d^* is estimated from

$$(11.4) \qquad s_{d^*} = \sqrt{\frac{s_1^2}{n_1} + \frac{s_2^2}{n_2}}$$

and the critical value, d_c^*, is obtained from

$$(11.5) \qquad d_c^* = 0 \pm z s_{d^*}$$

Assume a group of female employees brings suit against a large manufacturing company, alleging discrimination on the basis of sex. Specifically, the group claims females are routinely discriminated against in payment of wages and salaries. To determine if there is basis for the suit, attorneys for the plaintiffs hire a statistician to sample randomly the records of employees by sex within a certain pay classification. Using computer printouts, the statistician randomly selects 100 male (sample 1) and 100 female (sample 2) employees, resulting in the following values:

$$\bar{x}_1 = \$14{,}500 \quad s_1 = 2{,}250 \quad n_1 = 100$$

$$\bar{x}_2 = \$13{,}200 \quad s_2 = 2{,}050 \quad n_2 = 100$$

Both populations and population standard deviations are unknown; population standard deviations are estimated by the sample standard deviations. The test statistic is d^*, while the sampling distribution of d^* is assumed to be normal. The standard error, d^*, is estimated by use of Equation 11.4:

$$s_{d^*} = \sqrt{\frac{s_1^2}{n_1} + \frac{s_2^2}{n_2}} = \sqrt{\frac{(2250)^2}{100} + \frac{(2050)^2}{100}}$$

$$s_{d^*} = 304.38$$

In testing the difference between earnings of males and females, assume the statistician uses the .05 level of significance. The hypotheses are

$$H_0: \quad \mu_1 = \mu_2$$
$$H_1: \quad \mu_1 \neq \mu_2$$

and critical values of d^* (Equation 11.5) are:

$$d_c^* = 0 \pm zs_{d^*}$$

$$d_c^* = 0 \pm 1.96\ (304.38)$$

$$d_c^* = -\$596.58 \text{ and } +\$596.58$$

The decision rule in this situation is:

Random samples of $n_1 = 100$ *and* $n_2 = 100$ *are drawn from each population. The mean of each sample is calculated and the value of* $d^* = \bar{x}_1 - \bar{x}_2$ *is computed. If* d^* *is less than or equal to* $-\$596.58$ *or greater than or equal to* $+\$596.58$, H_0 *is rejected. Otherwise,* H_0 *is accepted.*

One additional calculation, d^*, is required to determine if the difference is significant. The value of this statistic is

$$d^* = \$14{,}500 - \$13{,}200 = \$1{,}300$$

Since d^* is greater than \$596.58, H_0 is rejected. Thus, the statistician for the plaintiffs concludes that males in the company earn more than females in the pay classification sampled. The test is illustrated in Figure 11.5.

Two Sample Means—Small Samples

Another kind of statistical problem which finds many examples in the real world is testing whether the means of two small samples are significantly different. Usually, small samples taken from a population have larger standard errors than do large samples. If an extremely large or extremely small value is chosen at random from the population, the influence of such a value on the standard error of a small sample is greater than on the standard error of a large sample. Because a small sample has a larger standard error, any estimates about a population parameter are less precise than from large samples. Of course, if the population

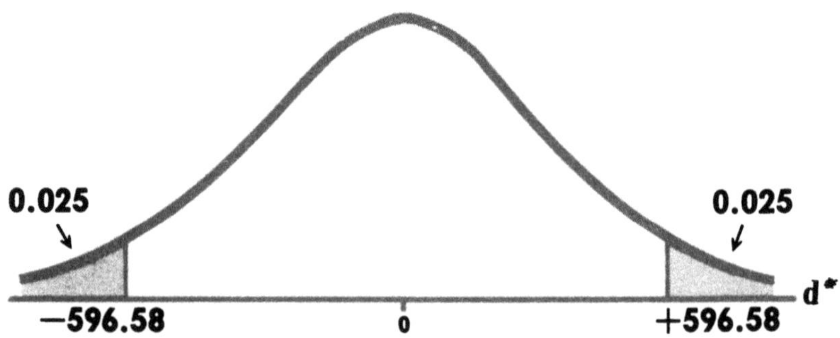

Figure 11.5
Test of Difference between Mean Earnings of Male and Female Employees

standard deviation is known, these sampling fluctuations are nonexistent, for it is not necessary to estimate σ by computing the standard deviation. This is seldom the case. It is unreasonable to expect that the standard deviations of the two populations would be known. For the fundamental purpose of sampling is to estimate a population parameter, such as a mean or standard deviation, which is unknown.

Compensation must be made for a larger error involving small samples. As noted earlier, student's t is the appropriate test statistic for problems involving small samples. The procedure used to test hypotheses involving a difference in the means of small samples is quite similar to that developed in the preceding section. There are, however, three assumptions involved: (1) the populations are normally distributed; (2) the population variances are equal; and (3) the two samples are independent. If the assumption regarding variances is valid, then an estimate of the common population standard deviation is made by *pooling* the variances from both samples as follows:

$$(11.6) \qquad s_p = \sqrt{\frac{(n_1 - 1)s_1^2 + (n_2 - 1)s_2^2}{n_1 + n_2 - 2}}$$

where

s_p = pooled estimate of the common population standard deviation
$n_1 - 1$ = degrees of freedom in sample 1
$n_2 - 1$ = degrees of freedom in sample 2
s_1^2 = variance of sample 1
s_2^2 = variance of sample 2

The pooled estimate, s_p, is used to estimate the standard error of d^*:

$$(11.7) \qquad s_{d^*} = \sqrt{\frac{s_p^2}{n_1} + \frac{s_p^2}{n_2}}$$

Or, more simply, the standard error of d^* is calculated by

$$(11.8) \qquad s_{d^*} = s_p\sqrt{\frac{1}{n_1} + \frac{1}{n_2}}$$

Critical values, d^*, are based on the t statistic and s_{d^*}. As noted earlier, the sampling distribution of d^* must be treated as a t distribution because the population standard deviations are unknown—analogous to treatment of the sampling distribution of $\bar{x}$ as a t distribution in single-sample tests when the population standard deviation is unknown. The degrees of freedom for a two-sample test are $n_1 + n_2 - 2$; the critical values of d^* are obtained from

$$(11.9) \qquad d_c^* = 0 \pm t s_{d^*}$$

If the degrees of freedom from the two-sample test of means exceed 30, the z statistic may be used.

To illustrate, consider the sex discrimination case discussed earlier. Now assume, in the interest of time, that the statistician retained by the plaintiffs' attorneys randomly samples the records of 15 male and 15 female employees within the given pay classification. The following values result from his samples:

$$\bar{x}_1 = \$14{,}100 \quad s_1 = \$2{,}900 \quad n_1 = 15$$

$$\bar{x}_2 = \$12{,}900 \quad s_2 = \$2{,}580 \quad n_2 = 15$$

The pooled estimate of the common population standard deviation, s_p, is estimated by using Equation 11.6:

$$s_p = \sqrt{\frac{(n_1 - 1)s_1^2 + (n_2 - 1)s_2^2}{n_1 + n_2 - 2}}$$

$$= \sqrt{\frac{(14)(2{,}900)^2 + (14)(2{,}580)^2}{15 + 15 - 2}}$$

$$= \$2{,}744.67$$

In the above situation, the test statistic is $d^* = \bar{x}_1 - \bar{x}_2$. With sample sizes of $n_1 = 15$ and $n_2 = 15$, there are 28 degrees of freedom. The sampling distribution of d^* is normal, but critical values are based on the t distribution. The standard error of d^* is estimated by using Equation 11.8:

$$s_{d^*} = s_p \sqrt{\frac{1}{n_1} + \frac{1}{n_2}}$$

$$= \$2{,}744.67 \sqrt{\frac{1}{15} + \frac{1}{15}}$$

$$= \$1{,}002.21$$

To test the difference between the earnings of males and females, the statistician uses the .05 level of significance. The hypotheses are

$$H_0: \quad \mu_1 = \mu_2$$
$$H_1: \quad \mu_1 \neq \mu_2$$

and the critical values of d^* (Equation 11.9) are:

$$d_c^* = 0 \pm t s_{d^*}$$

$$d_c^* = 0 \pm (2.048)\ (1002.21)$$

$$d_c^* = -\ \$2{,}052.53 \text{ and } +\$2{,}052.53$$

One additional calculation, d^*, is required to determine if the difference between mean earnings of males and females is significant. The value of this statistic is $d^* = \$14,100 - \$12,900 = \$1,200$. Since d^* is less than \$2,052.53, H_0 is accepted. Based on this sample, the statistician for the plaintiffs concludes that there is no difference in the earnings of males and females. Any difference is due to chance.

TESTING DIFFERENCES BETWEEN PROPORTIONS

Population Proportion and Sample Proportion

A common test of significance in decision making is to compare the value of a proportion derived from a sample with the value of a known population proportion. Since the sampling distribution of a sample proportion is binomial, the exact test of significance involves evaluation of the probabilities of obtaining certain values of the binomial distribution. Therefore, it might not be possible to design a test of the population proportion having exactly the significance level desired because of the discreteness of the distribution. This is especially true when the sample size is small. When samples are large, the binomial distribution may be approximated by a normal distribution.

The procedure for testing hypotheses about proportions is identical to that followed for tests of means. Therefore, discussion of testing of a hypothesis concerning a known population proportion will be rather brief.

To illustrate testing of proportions, consider the following example. Based on historical evidence, a candidate for the U.S. Senate must poll 75 percent of the vote in the southeastern part of a state to be elected. The incumbent wants to assess his chances of remaining in office in the next election. He retains a polling firm to conduct a poll in the southeastern section. A survey is conducted in the southeastern part of the state; 900 of the 1,500 registered voters selected at random indicate they favor the incumbent candidate.

The testing procedure is essentially the same as testing the difference between the population mean and the sample mean. The z test is the appropriate test for the difference between a sample proportion and a known population proportion. If π represents the assumed population proportion and p the sample proportion, the critical value of z is given by the following:

$$(11.10) \quad z = \frac{p - \pi}{\sigma_p}$$

where

π = population proportion
p = sample proportion
σ_p = standard error of the population proportion

Remember from Chapter 9 that the standard error of the population proportion was calculated by Equation 9.9:

$$\sigma_p = \sqrt{\frac{\pi(1 - \pi)}{n}} = \sqrt{\frac{(.75)(.25)}{1500}} = .011$$

In this example, the hypothesis to be tested is that the population proportion (π) is equal to or greater than .75. This is the same as saying that the proportion of persons favoring the incumbent is the population proportion of .75. The null hypothesis is that the proportion of voters favoring his return to office is not significantly less than $\pi = .75$. The alternate is that the proportion of voters favoring the incumbent is less than .75. The hypotheses are

$$H_0: \quad \pi \geq .75$$
$$H_1: \quad \pi < .75$$

and, assuming $\alpha = .05$, the critical value of z using Equation 11.10 is:

$$z = \frac{p - \pi}{\sigma_p} = \frac{.60 - .75}{.011} = -13.636$$

Since the hypothesis is tested in only one direction, a one-tailed test is appropriate. A one-tailed test is used because the incumbent senator is concerned only if the sample proportion is less than the population proportion of .75. The tabular value of z at the .05 level of significance is 1.645. Thus, the computed value of z falls in the rejection region, and H_0 is rejected. The difference of 15 percentage points between the sample proportion and that necessary to carry the state is significant. Based on the results of this sample, one would conclude that it appears unlikely the incumbent senator will be reelected.

Two Sample Proportions

Comparison of two sample proportions to determine whether there is significant difference is similar to the test of difference between the means of two samples. The probabilities associated with the normal curve may be used as an approximate test of the difference between two proportions. When the samples are reasonably large, this approximation is sufficiently accurate and is almost universally used.

The task of determining whether two populations differ with respect to a certain attribute occurs rather frequently in various types of demographic analyses—e.g., public opinion polls and market research—as well as in many types of industrial applications. For example, is there any difference in the television viewing habits of men and women? Do smoking habits vary between males and females? Is there a difference in the percentage of defective parts produced by two machines on an assembly line? Or is there a difference in percentage of registered voters of people of different races?

Problems of this type can be treated by testing the null hypothesis H_0: $\pi_1 = \pi_2$. The alternate hypothesis is H_1: $\pi_1 \neq \pi_2$. The test is two-tailed, and the test statistic is $d^* = p_1 - p_2$.

Since p, the sample proportion, has a binomial distribution, a normal distribution having the same mean and standard deviation is used as an approximation of the binomial: Only those tests in which both samples are large enough for the sample proportions to be approximately normally distributed will be considered. Therefore, if p_1 has a normal distribution and p_2 has a normal distribution, $d^* = p_1 - p_2$ will also have a normal distribution with a mean equal to $\pi_1 - \pi_2$. The standard error of d^* is

$$(11.11) \quad \sigma_{d^*} = \sqrt{\frac{\pi_1(1 - \pi_1)}{n_1} + \frac{\pi_2(1 - \pi_2)}{n_2}}$$

Because the values of π_1 and π_2 are unknown, they are estimated by using sample values. Although the parameters are unknown, they are assumed to be equal under the hypothesis H_0: $\pi_1 = \pi_2$. If this is the case, then there is a single value of π which is common to both populations. In order to evaluate σ_{d^*} from sample data, it is necessary to estimate the common value of π by pooling the data from both samples. Designated $\hat{p}$, the pooled estimate of π is

$$(11.12) \quad \hat{p} = \frac{n_1 p_1 + n_2 p_2}{n_1 + n_2}$$

The pooled estimate is substituted for π to obtain s_{d^*}, the sample estimate of σ_{d^*}. Thus,

$$(11.13) \quad s_{d^*} = \sqrt{\frac{p(1 - p)}{n_1} + \frac{p(1 - p)}{n_2}}$$

Critical values of d^* are then obtained from

$$(11.14) \quad d_c^* = 0 \pm z s_{d^*}$$

To illustrate testing of differences between two sample proportions, suppose a company has over several years hired assembly line workers on the basis of a written test. The test is designed to measure general intelligence and personality characteristics desirable for assembly line workers. No work experience is necessary. In order to be hired, an applicant must successfully complete the test.

A black male applies for a job, takes the test, and scores below the passing level of 70. After exhausting administrative remedies, he sues under Title VII on the grounds that the test has a discriminatory impact on blacks. The complaint alleges that, over a period of years, 240 out of 400 white applicants passed the test while only 165 out of 300 black applicants were successful.

The defendant company retains a law firm to represent it. One of the first questions to be answered relates to the significance of the difference between scores of blacks and whites. That is, is there a significant difference between the proportion of whites who passed the test as contrasted with the proportion of blacks? If the population of whites is designated population 1 and the population of blacks population 2, then

$$p_1 = \frac{240}{400} = .60$$

and

$$p_2 = \frac{165}{300} = .55$$

The pooled estimate of the common population proportion, π (Equation 11.12), is

$$\begin{aligned}\hat{p} &= \frac{n_1p_1 + n_2p_2}{n_1 + n_2}\\ &= \frac{(400)(.60) + (300)(.55)}{400 + 300}\\ &= .579\end{aligned}$$

The standard error of d^* is estimated by use of Equation 11.13:

$$\begin{aligned}s_{d^*} &= \sqrt{\frac{\hat{p}(1-p)}{n_1} + \frac{\hat{p}(1-p)}{n_2}}\\ &= \sqrt{\frac{(.579)(.421)}{400} + \frac{(.579)(.421)}{300}}\\ &= .03771\end{aligned}$$

To test the difference between the proportions of whites and blacks passing the employment test, assuming a .05 level of significance, the hypotheses are

$$\begin{aligned}H_0&: \quad \pi_1 = \pi_2\\ H_1&: \quad \pi_1 \neq \pi_2\end{aligned}$$

and the critical values of d^* (Equation 11.14) are:

$$\begin{aligned}d_c^* &= 0 \pm zs_{d^*}\\ &= 0 \pm (1.96)(.03771)\\ &= -\ .0739 \text{ and } +.0739\end{aligned}$$

One additional calculation, d^*, is required to determine whether the difference between the proportion of whites and blacks passing the test is significant. The value of the statistic is $d^* = p_1 - p_2 = .60 - .55 = .05$. Since d^* is less than .0739, the null hypothesis is accepted. Based on this sample, one would conclude that there is no significant difference in the proportion of whites and blacks who pass the employment test.

ILLUSTRATIVE USES OF STATISTICAL INFERENCE

Numerous cases have utilized statistical inference. One of the better examples is *Chance v. Board of Examiners,*[1] a case decided by the U.S. district court and affirmed by the U.S. court of appeals.[2] At both levels, statistics and statistical techniques are mentioned prominently. In this case, a class action in which the question was raised as to the constitutionality of examinations required for appointment to supervising positions in the city's school system, the court found for the plaintiff and issued an injunction against conducting such examinations.

The plaintiff offered the testimony of Dr. Jacob Cohen, a statistician, while the board adduced the testimony of Dr. Nathan Jasper, also a statistician. Cohen presented data to show that the examination administered by the board discriminated against black and Puerto Rican candidates. He presented a survey which revealed that out of 6,201 candidates taking most of the supervisory examinations during the last seven years, 5,910 were identified by race; of the 5,910, 818 were black or Puerto Rican and 5,092 were Caucasian. Cohen noted that 31.4 percent of the 818 black and Puerto Rican candidates passed, compared with 44.3 percent of the white candidates. Using the data presented in Table 11.2, the court noted:

> These overall figures, however, tell only part of the story. Of greater significance are the results of two examinations which had by far the largest number of candidates, those for Assistant Principal of Day Elementary School and Assistant Principal of Junior High School. . . . Thus white candidates passed the examination for Assistant Principal of Junior High at almost double the rate of Black and Puerto Rican candidates, and passed the examination for Assistant Principal of Day Elementary School at a rate one-third greater than Black and Puerto Rican candidates. The gross disparity in passing rates on these two examinations is of particular significance.[3]

As will be noted from Table 11.2, Cohen obviously tested the difference in the two proportions and found the differences to be highly significant. He concluded that the difference was due to other than chance—namely, the discriminatory nature of the examinations with respect to blacks and Puerto Ricans.

A similar case, *Boston Chapter, NAACP, Inc., v. Beecher,*[4] relied heavily on the use of differences in proportions. Plaintiff presented the combined percentage of black-Spanish applicants who passed the test for firefighter positions as contrasted to the percentage of whites passing the exam. These statistics showed that 55 percent of the white applicants taking the exam passed, while only 39 percent

Table 11.2
Results of Examinations for Assistant Principal,
Day Elementary and Junior High Schools, by Race, New York

Examination	Caucasian	Black	Puerto Rican	Black and Puerto Rican	Probability of Chance Result Less Than
Assistant principal, day elementary schools	1171 (61.3)	278 (45.68)	7 (28.57)	285 (45.26)	1/1 million
Assistant principal, junior high schools	1319 (48.82)	236 (26.27)	14 (14.29)	250 (25.60)	1/1 million

SOURCE: *Chance v. Board of Examiners*, 330 F. Supp. 210 (1971).

NOTE: Numbers in parentheses are percentages.

of the minority group taking the exam passed. The court held, however, that these data were not sufficient to establish a prima facie:

> The Court is not impressed by such obviously meager statistics although it does recognize that some courts have indeed relied on such unconvincing statistics. . . . This Court does not find the available exam statistics sufficient in themselves to establish a prima facie showing that the exam is discriminatory.[5]

In describing the "meager statistics," the court was referring to the fact that the sample consisted of 3,089 whites, 18 blacks, and 15 of Spanish origin. However, by combining the exam data with employment and populations data from the city of Boston, which had a combined minority population of 23 percent but only a .9 percent minority representation on the firefighter force, the court ruled that a prima facie case was established.

Significance level is addressed in a straightforward manner in *United States v. Georgia Power Company,*[6] a case which has been cited frequently in this book. Referring to a guideline which required it to be shown that there was no more than 1 chance out of 20—i.e., the .05 level—that job relatedness occurred by coincidence, the court quoted from the brief filed by the American Psychological Association:

> In short, the 5% level is an arbitrary convention. It has never been a part of professional standards to insist on a specific level of confidence; however, when one departs from the conventional standard, he should ordinarily have a reason which can be stated in advance of the analysis. Under some circumstances, for example, one might need a prediction so badly or have such severely restricted range of scores on prediction or criterion that he will decide to accept any finding significant at the 10% level; under other circumstances, the cost of using a test that might later prove invalid may be so great that the investigator will insist on at least a 1% level.[7]

One additional example will suffice. This illustrates the application of statistical inference in determining economic losses encountered by a vertically integrated poultry firm. A vertical tier of egg production and marketing processes was under the ownership and control of one firm. Vertical integration involves the operation of a single firm at more than one stage of production, which is common in industrial processes; its application has grown in agriculture.

The problem demonstrated itself through reduced levels of egg production in all layer flocks controlled by the firm. Production levels among the 25 flocks, comprising about 500,000 layers, ranged from 9.9 dozen to 25 dozen eggs per year with a mean of 17.88 dozen and a standard deviation of 1.73 dozen. Extensive investigation of the possible causes of reduced production led to the breeder flock. Eggs produced by the breeder flock were the source of birds in the layer flocks. It was hypothesized that a genetic fault existed in the breeder flock at the time it was purchased by the firm. With the problem supposedly originating in the breeder flock, the first tier, and not significantly revealing itself until the egg production stage, all tiers of the production processes controlled by the firm were affected. Thus suit was brought against the breeder to recover damages sustained because of the lower level of egg production.

The initial role of the statistician in this case was to determine whether the difference between the expected production per layer of 19 dozen eggs per year (the national norm) and the average annual output of the flock in question of 17.88 dozen could have occurred due to chance. Using the *t*-test for differences between the population mean of 19 and the sample mean of 17.88, the difference was found to be significant at the .005 level. That is, the probability is less than .005 (1 in 200) that a difference occurring by chance supported the firm's position that genetic factors contributed to reduced egg output.

Lost production was defined as the difference between the actual lay and the expected norm of 19 dozen eggs per layer per year. Once the extent of production loss was calculated, appropriate egg prices existing at the time were used in determining the value of the lost production. This was used as the basis for an out-of-court settlement in this case.

NOTES

1. 330 F. Supp. 203 (1971).
2. 458 F. 2d 167 (1972).
3. 330 F. Supp. 210 (1971).
4. 371 F. Supp. 507 (1974).
5. *Id* at 514.
6. 474 F. 2d 906 (1973).
7. *Id* at 915.

SELECTED REFERENCES

Arnold, B. C. "Hypothesis Testing with a Preliminary Test of Significance." *Journal of the American Statistical Association,* Vol. 65, No. 332 (December 1970), pp. 1590–96.

Braun, Louis J. "Statistics and the Law: Hypothesis Testing and Its Application to Title VII Cases." *The Hastings Law Journal,* Vol. 32 (September 1980), pp. 59–88.

Freund, John E. *Mathematical Statistics,* Chaps. 11–12. 2d ed. Englewood Cliffs, N.J.: Prentice-Hall, 1971.

Lehman, E. L. *Testing Statistical Hypotheses.* New York: John Wiley & Sons, 1959.

Lindgren, B. W. *Statistical Theory,* Chap. 7. New York: Macmillan Co., 1962.

Yamane, Taro. *Statistics: An Introductory Analysis,* Chap. 9. 3d ed. New York: Harper & Row, 1973.

CHAPTER 12

THE CHI-SQUARE DISTRIBUTION

In the previous chapters on statistical inference, it was often necessary to make certain assumptions about the populations from which the samples were drawn. In many of the statistical tests, for example, it was assumed that the samples came from normally distributed populations. Furthermore, the validity of the samples was found to be dependent upon one's ability to estimate the mean and standard deviation of the population. The added assumption of equal variances was required in other tests involving two or more samples. Often, however, these assumptions cannot be justified; it is necessary to use procedures that do not require that these assumptions be met. These procedures are generally referred to as *nonparametric* methods.

A nonparametric or distribution-free test is one that does not require knowledge of any parameter of the population. In general, nonparametric tests involve simple computations and can be applied under very few restrictions. In this chapter, a statistic known as chi-square, χ^2, is discussed; this is often called the χ^2 analysis of frequencies. Two different types of χ^2 tests will be discussed:

1. Testing for significant difference between observed and expected frequencies in a single-variable relationship.

2. Testing to determine whether two different qualities or characteristics of a population are independent or *contingent* upon one another.

THE CHI-SQUARE DISTRIBUTION

The sampling distribution of χ^2 is called the chi-square distribution with a single parameter, degrees of freedom. Introduced in 1900 by Karl Pearson, chi-square is the most important of all distribution-free tests. While the test is distribution-free, since it makes no assumption about the population from which the sample is drawn, there is a rigidly defined chi-square distribution. The chi-square distribution has zero as its lower limit and extends to infinity in the positive direction. Therefore, negative values of χ^2 are impossible. The distribution is skewed to the right (Figure 12.1). Values of χ^2 for certain levels of

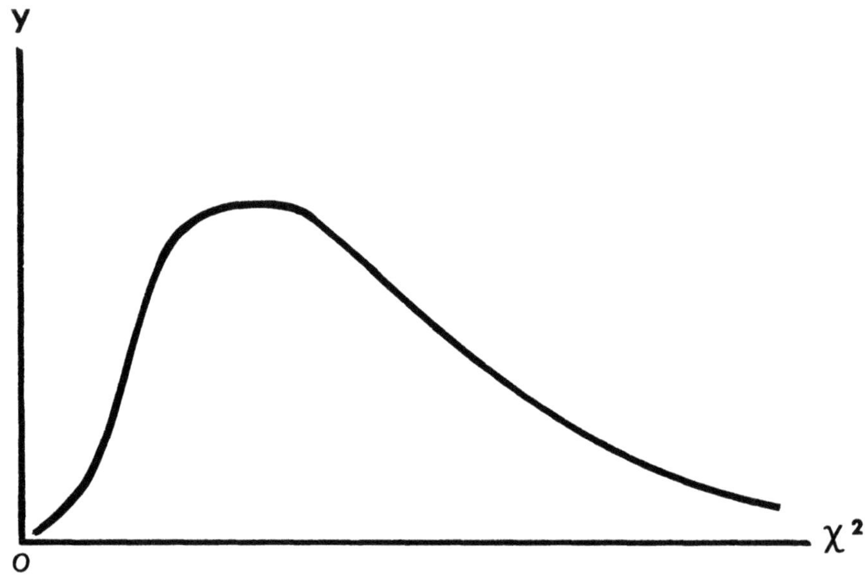

Figure 12.1
Typical Chi-Square Distribution

significance, α, are shown in Appendix Table A.4. Degrees of freedom appear in the extreme lefthand column of the chi-square distribution. The proportions of the area in the right tail are specified in the other column headings.

CHI-SQUARE TEST—SAMPLING FREQUENCIES AGAINST PRESUPPOSED FREQUENCIES

Chi-square tests deal with frequencies against presupposed frequencies or counts, i.e., the number of observations that fall into certain categories. In each test, there is a set of frequencies obtained from a random sample. The statistic χ^2 as defined by Equation 12.1 is used to test for significant difference between observed and expected frequencies of n events, where $n > 2$:

$$\chi^2 = \sum_{i=1}^{n} \frac{(O_i - E_i)^2}{E_i} \qquad (12.1)$$

where

O_i = observed frequency of the ith event
E_i = expected frequency of the ith event

As shown in Equation 12.1, the chi-square statistic compares the differences between two sets of frequencies—observed and expected—and expresses them with a single numerical value. If these differences, as represented by the value χ^2, are significant at the level selected for the test, the null hypothesis is rejected. Critical values of χ^2 are obtained from the tables of the chi-square distribution.

To illustrate, assume a media consulting group surveys the radio audience in a certain city to determine listenership by radio station. Because the survey of 555 listeners indicates a pattern at variance with past studies of a similar nature, several stations bring suit against the consulting group. Plaintiffs' attorneys retain a statistician to evaluate the results of the survey.

As part of the study, the consulting group publishes listenership by demographics, i.e., by income, age, sex, and race. To check the validity of the study, particularly the representativeness and randomness of the sample, the statistician compares survey results by income with expected income levels in the population as reported by the U.S. Commerce Department. The observed and expected frequencies and the calculation of χ^2 are summarized in Table 12.1.

The null and alternate hypotheses are:

H_0: Survey distribution of income is identical to actual distribution of income
H_1: Survey distribution of income is not identical to actual distribution of income

The degrees of freedom in this test are one less than the number of cells, or six. In general, for tests involving a single variable of classification, the degrees of freedom are equal to the number of cells minus one. The critical value of χ^2 with six degrees of freedom at the .05 level of significance is obtained from the table of the chi-square distribution in Table A.4. This value is 12.59. Since the calculated value of χ^2 is greater than the critical value, the null hypothesis is rejected and the alternate is accepted. Based on the income data reported, the

Table 12.1
Observed and Expected Levels of Income, Listenership Survey, January 1983

Income Level (Cells)	Observed Frequency (O)	Expected Frequency (E)	$\frac{(O-E)^2}{E}$
Less than $10,000	100	85	2.647
$10,000 but less than $20,000	110	120	.833
$20,000 but less than $30,000	105	110	.227
$30,000 but less than $50,000	90	125	9.800
$50,000 but less than $75,000	80	70	1.429
$75,000 but less than $100,000	40	30	3.333
Over $100,000	30	15	15.000
		χ^2 =	33.269

statistician concludes that the income reported in the survey is not equal to the income for the geographical area as reported by the Commerce Department. Such a finding casts doubts on the representativeness of the sample and, hence, on the validity of the listenership study.

CONTINGENCY TABLES

The chi-square statistic can be used to test hypotheses in many situations. In the previous section, χ^2 was applied to one variable of classification. In this section, it will be used with two variables of classification. For example, radio listeners can be classified according to listenership and income, listenership and sex, listenership and age, and the like. When observations are classified according to two variables and arranged in a table, the display is called a *contingency table*. One of the variables is entered in columns, the other in rows.

Given the contingency table, the chi-square statistic can be used to test whether the two variables of classification are independent of one another. If two classification variables are independent, then membership in a particular class of one variable has no effect on membership in a particular class of the other variable. In the examples cited earlier, if sex and listenership are independent, then a female is just as likely to listen to one radio station as another. Or, if income and listenership are independent, a person earning $10,000 is just as likely to be listening to a particular station as one earning $100,000. That is, income level does not influence listenership to a station.

In tests of independence, the null hypothesis is always that the two classification variables are independent. The alternate hypothesis is that they are not independent. The chi-square statistic is computed as follows:

$$(12.2) \qquad \chi^2 = \sum_j \sum_i \frac{(O_{ij} - E_{ij})^2}{E_i}$$

where

O_{ij} = observed value in ith row and jth column
E_{ij} = expected value in ith row and jth column

The procedure can perhaps be best explained through examples. A statistician was recently employed by attorneys for the defendant in a criminal trial. The purpose of his employment was to survey individuals in the area in which the trial was to be held to assist the defense team in selecting jurors. A random telephone survey was conducted of 400 adults living in Houston County, the scene of the trial. Calls, based on randomly generated numbers, were made to various areas in Houston County, the number of calls to each area being proportional to the relative number of telephone listings in the area. A total of 20 questions was asked of each respondent.

Several questions were structured to ascertain whether the defendant's name was associated with the crime he allegedly committed. One question related to association of the defendant's name by sex of the respondent; it is summarized in Table 12.2. The purpose here was to determine whether name association was independent of sex. Such information would be valuable in jury selection.

The hypotheses to be tested at the .01 level of significance are:

H_0: Sex of respondent and name association are independent
H_1: Sex of respondent and name association are not independent

To determine expected frequencies, it is assumed that the two classification variables are independent. If this is true, one would expect the proportion of "yes" responses and the proportion of "no" responses to be the same for both sexes. The expected frequency of any cell, E_{ij}, is the product of the total of the *i*th row and the *j*th column, divided by *n*, the sample size. That is,

$$E_{ij} = \frac{t_i t_j}{n} \tag{12.3}$$

where

t_i = total of the *i*th row
t_j = total of the *j*th column

For example, the expected number of "yes" responses for males would be

$$E_{ij} = \frac{(202)(141)}{376} = 82.5$$

Expected values for the other cells in the table are calculated in a like manner:

Response	*Male*	*Female*
Yes	82.5	137.5
No	58.5	97.5

Table 12.2
Cross-Tabulation of Responses,
by Sex of Respondent, January 1983

Response	**Sex of Respondent** *Male*	*Female*	**Total**
Yes	99	121	220
No	42	114	156
Total	141	235	376

As in a single-variable relationship, the calculated value of χ^2 must be compared with a critical value for the significance level selected with the appropriate number of degrees of freedom. The degrees of freedom for a chi-square test of independence will always be $(r - 1)(c - 1)$, where r represents the number of rows and c the number of columns in the contingency table. For this test of independence between sex and name association, the number of degrees of freedom is $(2 - 1)(2 - 1) = 1$.

When there are only two rows and two columns in a contingency table, chi-square has only one degree of freedom. Under these circumstances, an adjustment called the *Yates correction factor* should be used. The correction is accomplished by subtracting .5 from the absolute value of the difference between O and E before squaring. Thus, the corrected value of χ^2 is obtained from

$$\chi^2 = \sum \frac{(|O - E| - .5)^2}{E} \tag{12.4}$$

The calculations for the test of independence between sex of respondent and name association of the defendant are presented in Table 12.3.

The critical value of chi-square with one degree of freedom at $\alpha = .01$ is 6.63. Since the calculated value of 11.967 is greater than the critical value, H_0 is rejected. It can be concluded that sex of respondent and association of the name of the defendant with the crime are not independent. Based on the results of the survey, males were much more likely to associate the defendant's name with the crime than were females.

Another question in the survey related to association of the defendant's name with the crime on the basis of the educational level of respondents. That is, responses were cross-tabulated by educational level of the respondents (Table 12.4).

The calculations for the test of independence between educational level and association of the defendant's name with the crime are shown in Table 12.5.

Since there are two rows and six columns in the contingency table, there are five degrees of freedom for the chi-square test of independence. At the .01 level

Table 12.3
Test of Independence, Sex versus Name Association, January 1983

Response	O	E	$(\lvert O - E\rvert - .5)^2/E$
Yes—Male	99	82.5	3.103
Yes—Female	121	137.5	1.862
No—Male	42	58.5	4.376
No—Female	114	97.5	2.626
	376	376.0	$\chi^2 = 11.967$

Table 12.4
Cross-Tabulation of Responses, by Educational Level of Respondent, January 1983

	Educational Level						
Response	*Less Than High School*	*High School Graduate*	*Some College*	*College Graduate*	*Some Post Degree*	*Post Degree*	**Total**
Yes	20	77	62	58	7	8	232
No	55	60	36	9	1	1	162
Total	75	137	98	67	8	9	394

of significance, the critical value of chi-square is 15.09. Since the calculated value of 60.776 is greater than the critical value, the null hypothesis of independence is rejected and the alternate is accepted. Therefore, it can be concluded that educational level of respondents and association of the defendant's name with the crime are not independent. Results of the survey indicate that those individuals with less than a high school education were less likely to associate the defendant's name with the crime than were individuals of any other educational level. This finding, coupled with the fact that males appear to be much more likely to associate the defendant's name with the crime than were females, should be of considerable assistance to the defense attorneys.

Table 12.5
Test of Independence, Educational Level versus Name Association, January 1983

Response	*O*	*E*	$\frac{(O-E)^2}{E}$
Yes—Less than high school	20	44.2	13.250
Yes—High school graduate	77	80.7	.170
Yes—Some college	62	57.7	.320
Yes—College graduate	58	39.4	8.781
Yes—Some post degree	7	4.7	1.126
Yes—Post degree	8	5.3	1.375
No—Less than high school	55	30.8	19.014
No—High school	60	56.3	.243
No—Some college	36	40.4	.479
No—College graduate	9	27.5	12.445
No—Some post degree	1	3.3	1.603
No—Post degree	1	3.7	1.970
Total	394	394.0	$\chi^2 = 60.776$

ILLUSTRATIVE USES OF CHI-SQUARE

Chi-square was used in a recent case involving procedures employed by a film distributor to validate box office receipts reported by a film exhibitor, Pike Theater. Distributors of films license motion pictures to exhibitors on a "flat" rental or on a "percentage" basis. The flat rental system specifies a sum to be paid the distributor by the exhibitor regardless of the film's success; on the percentage basis, rental is calculated as a fraction of the gross admission receipts at the box office. In this lawsuit, the percentage basis was used; the critical factor in ascertaining accurate gross box office receipts was the number of patrons as established by sale of admission tickets.

Distributors commonly employ two systems to verify box office receipts reported by exhibitors. When a check of the box office is done with the knowledge of the exhibitor, it is referred to as an *open check*. When done covertly, it is called a *blind check*. The open check is conducted by an inspector who identifies himself to the theater manager and informs him that he is there to check receipts for that particular date. In the conventional or four-wall theater, such as Pike Theater, the checker usually positions himself so as to be able to count the patrons. In the blind check, the same procedure is used except that the theater manager does not know he is being audited.

Based on five blind audits of Pike Theater, the film distributor alleged underreporting of box office receipts and brought suit to recover damages. At this point, attorneys for the defendant, Pike Theater, employed a statistician to examine the procedure followed by the distributor in conducting the blind checks, which were conducted from July 1977 to March 1978. Five dates were included: July 8, July 9, March 10, March 15, and March 17. During this period, ten showings of three movies were checked. Three of the blind checks, involving six showings, were on Friday nights; the remaining two were on Wednesday and Saturday.

In his analysis, the statistician initially addressed the concept of sampling theory, noting that there is no best method of selecting or drawing a sample. He pointed out, however, that samples must be drawn so as to reflect the special conditions of the population and under certain ground rules in order to be able to use the results—to be able to infer from the sample to the population. Clearly, the reliability of the conclusions drawn concerning a population depends on whether the sample is properly chosen so as to represent the population sufficiently well.

Examining the procedure followed by the distributor, in addition to addressing the inadequacy of the sample size, the statistician questioned the sampling distribution by day of the week. There was a concentration of sampling on Friday nights and total exclusion of Sunday, Monday, Tuesday, and Thursday. One would question whether this could have occurred through chance or whether there was possible bias in selection of sampling days. For example, during the four days in which no showings were sampled, one would have expected a total of 5.62 blind checks had the days been randomly selected. Conversely, the expected blind checks on Fridays would have been 1.28 (Table 12.6).

Based on the data in Table 12.6, the hypothesis was tested that the sample was randomly selected; the alternate was that the sample was not randomly selected. Using a significance level of .05, the critical value of χ^2 with six degrees of freedom is 12.59. Since the calculated value of χ^2 exceeds the critical or tabular value, H_0 was rejected and H_1 accepted. The statistician concluded that the sample was not randomly selected relative to days of the week.

The statistician also examined the sampling procedure on the basis of the months in which samples were drawn. Since the time period involving the blind check covered some nine months, one would expect a representative sample to include several of these months. However, only two months, July and March, were included in the sample; approximately 15.8 percent of the population of movie showings was sampled in March and 7.1 percent in July. Could this situation have occurred by chance? Expected observations for the seven months August through February were 8.38; in reality, there were no observations during this time period. Likewise, with only 38 total showings in March, expected showings observed would have been .657 had the sample been randomly selected. The actual number of showings observed was 6 (Table 12.7).

Using a significance level of .05 and with eight degrees of freedom, the critical value of χ^2 is 21.96. Since the calculated value of χ^2 of 61.307 exceeds the critical value, the null hypothesis is rejected and the alternate accepted. The conclusion is that the sample was not randomly selected in terms of the observed showing by month.

Based on his examination of the procedures followed by the film distributor in conducting blind checks of the theater, the statistician concluded that no statistical inferences could be drawn from the data. The film distributor ultimately withdrew the lawsuit against the defendant.

Table 12.6
Observed and Expected Showings, by Day of the Week and Number of Showings, Pike Theater, July 1977 to March 1978

Day of Week	Total Showings	Showings Sampled (O)	Expected Samples (E)	$\frac{(O - E)^2}{E}$
Sunday	108	0	1.868	1.868
Monday	72	0	1.246	1.246
Tuesday	72	0	1.246	1.246
Wednesday	72	2	1.246	.456
Thursday	72	0	1.246	1.246
Friday	72	6	1.280	17.405
Saturday	108	2	1.868	.009
Total	576	10	10.000	$\chi^2 = 23.476$

Table 12.7
Total and Expected Showings, by Month and Number of Showings, Pike Theater, January 1977 to March 1978

Month	Total Showings	Showings Sampled (O)	Expected Samples (E)	$\frac{(O - E)^2}{E}$
July	56	4	.969	9.481
August	70	0	1.211	1.211
September	68	0	1.177	1.177
October	72	0	1.246	1.246
November	68	0	1.177	1.177
December	71	0	1.228	1.228
January	71	0	1.228	1.228
February	64	0	1.107	1.107
March	38	6	.657	43.452
Total	578	10	10.000	$\chi^2 = 61.307$

A case in which chi-square played a noteworthy role is *Chance v. Board of Examiners,*[1] discussed in Chapter 11. Using the gross pass–fail statistics for persons who took examinations for supervisory positions in the New York City schools, Dr. Cohen, the statistician for the plaintiff, noted that the overall pass rate for white candidates (44.3 percent) was almost 1.5 times as high as the nonwhite rate (31.4 percent). He testified that ''on the basis of such a large sample (5,910 out of 6,201 candidates), the test results were especially valuable and formed a sound basis for drawing valid statistical conclusions as to the differences in passing rates between the ethnic groups involved.''[2] The court described Cohen's testimony:

In analyzing the statistics he used the chi-square test (Yates-corrected), which is a method using formulas generally accepted by statistical experts to determine whether an observed difference in any given sample is greater than that which would be expected on the basis of mere chance or probability. He found with respect to the aggregate test that by ''the chi-square (Yates-corrected) statistical test, the probability of the difference being a chance result not related to the factor of race is determined as less than one in one billion.''[3]

Defendant's expert witness, Dr. Jasper, attempted to rebut Dr. Cohen's testimony by showing that any attempt to use gross figures was inaccurate because of the possibility of ''overlap,'' i.e., the taking of more than one test by the same person. The court rejected Dr. Jasper's contention:

[I]f a ''random'' overlap existed . . . it would not substantially affect the probabilities. Dr. Jasper conceded, for instance, that a random overlap of as much as 50% would not substantially affect the significance of observed differences between the white and non-white races. Lastly, Dr. Cohen . . . testified that because of the large chi-square value an overlap would not significantly affect the probability data. In other words, the observed difference between the aggregate pass–fail rates was too great to be a mere matter of chance, unrelated to race . . . so that even if the probability were reduced somewhat because of overlap, the figure would still be significant.[4]

NOTES

1. 330 F. Supp. 203 (1971).
2. *Id* at 212.
3. *Id.*
4. *Id.*

SELECTED REFERENCES

Clark, Charles T. and Lawrence L. Schkade. *Statistical Analysis for Administrative Decisions,* Chap. 9. 3d ed. Cincinnati: South-Western Publishing Co., 1979.

Cochran, W. G. "The Chi-Square Test of Goodness of Fit." *Annals of Mathematical Statistics,* Vol. 23 (1952), pp. 315–45.

Conover, W. J. *Practical Nonparametric Statistics.* New York: John Wiley & Sons, 1971.

Gibbons, Jean D. *Nonparametric Statistical Inference.* New York: McGraw-Hill Book Co., 1971.

Guenther, William C. *Concepts of Statistical Inference,* Chap. 6, 2d ed. New York: McGraw-Hill Book Co., 1973.

Mann, H. B. "Nonparametric Tests Against Trend." *Econometrica,* Vol. 13 (1945), pp. 245–59.

Savage, I. R. "Bibliography of Nonparametric Statistics and Related Topics." *Journal of the American Statistical Association,* Vol. 48 (1953), pp. 844–906.

CHAPTER 13

CORRELATION AND REGRESSION

The process of observing two variables and deciding whether they vary together is a familiar experience. Much of human learning involves drawing conclusions based on associating the variation in one set of events with that of another. Mathematical expressions derived from this kind of analysis are frequently used in business, industrial, and scientific applications. For example, an economist may wish to forecast the demand for home computers by use of a mathematical expression that relates level of demand for computers with per capita personal income in various sales territories. The personnel director of a company has been advised to use test results to match personal characteristics with job requirements for the most effective placement of employees and to reduce his liability in case of a discrimination-in-hiring lawsuit. A drug company uses results that correlate the incidence of cancer in white mice with the dosage of a proposed new drug to determine whether the latter can be safely tested on human beings. In other studies, various physical laws have been verified by observing variables, analyzing data, and deriving mathematical expressions of the relationships between the variables.

The above examples illustrate the use of *correlation* and *regression*. The techniques involved in determining the relationship between a dependent variable and one or more independent variables are called *regression analysis*. Regression involves analysis of the variation of variables. The regression technique can be used (1) to derive a mathematical expression of the relationship between dependent and independent variables; (2) to provide information to use in deciding whether the variables are related; (3) to explain or account for the variations of the dependent variable; and (4) to furnish an equation which can be used to forecast or predict the values of the dependent variable.

Correlation analysis, a related technique, measures the strength of the relationship between variables. It concerns the analysis of the mutual variation of variables (1) to obtain a measure of their interrelatedness—called the correlation coefficient—and (2) to provide a basis for testing the degree or extent of association between the variables.

Regression and correlation—two powerful statistical tools—provide quantitative expressions or models which indicate the extent to which variables are related mathematically. It should be remembered that a cause-and-effect relationship should not necessarily be inferred. These statistical methods cannot offer proof of the existence of a relationship between selected variables; however, statistical analysis utilizing these methods can provide valuable information. This information can be used to support a judgment regarding the existence of a relationship between the variables being analyzed.

This chapter will consider linear regression only, i.e., regressions in which the relationship between the variables can be expressed by a linear equation. Since the primary thrust of the chapter is regression analysis, correlation will be treated as an adjunct to the regression analysis, even though correlation is frequently considered as a separate topic.

CORRELATION

As noted earlier, correlation is a measure of the degree of association between two variables. The correlation coefficient, r, will always be between -1 and $+1$. A value of $r = +1$ represents perfect positive correlation; $r = -1$ represents perfect negative correlation. When the two variables x and y have a correlation coefficient that approaches -1, this means that when x increases, y almost always decreases and vice versa. On the other hand, when x and y have a correlation coefficient close to $+1$, this means that they move predictably in the same direction. When x increases, y almost always increases and vice versa. If the correlation coefficient is close to 0, this means that no matter what direction x moves, y may move in any direction. That is, the two variables are not correlated; one cannot be used to predict the other.

Generally speaking, an r of $+.98$ would indicate a very high relationship between x and y; an r of $+.15$, a very low relationship. Likewise, an r of $-.98$ would indicate a very high relationship and $-.02$ very low association between x and y.

To illustrate, assume a law firm is retained to defend a firm which is being sued over the use of a test which is administered to prospective sales employees. A statistician is retained by the firm to determine whether the test is a valid predictor of how well the employee will function in the field. If a high degree of correlation exists between test scores and sales performance, this evidence can be used to help refute the plaintiff's allegation of discrimination in hiring.

The statistician randomly selects 20 sales employees from the personnel files of the company. He records test scores for these employees, along with records on sales performance expressed as the percentage of the time that each employee exceeded his or her weekly sales goal during the past year. The results are summarized in Table 13.1.

A quick method for determining whether there is an apparent relationship between two variables is to plot the points corresponding to each pair of values on a *scatter diagram*. A scatter diagram is simply a chart on which the values of

Table 13.1
Employee Test Scores and Employee Sales Performance, 20 Employees, January 1983

Employee	Employee Test Score (x)	Employee Sales Performance (y)
1	75	10
2	83	30
3	78	30
4	80	40
5	85	60
6	90	50
7	74	20
8	86	60
9	88	50
10	76	20
11	96	70
12	82	40
13	71	10
14	88	50
15	78	40
16	84	50
17	89	60
18	95	70
19	91	60
20	73	20

the independent variable, x_i, appear on the horizontal axis; the values of dependent variable, y, are plotted on the vertical axis. A scatter diagram for test scores and sales performance appears in Figure 13.1. Observing the pattern of points on the scatter diagram provides a rough indication of the extent and type of relationship, if any, between the two variables.

In Figure 13.1, there is an apparent relationship between x and y. Large values of x are apparently associated with large values of y, and small values of x with small values of y. The relationship appears to be a linear one, i.e., a straight line that slopes upward. Thus, the pattern tends to indicate the existence of positive linear correlation. Scatter diagrams, however, should be used simply as an indicator that a particular variable should be either eliminated as a possible predictor or investigated further.

After plotting and examining the scatter diagram, the *coefficient of correlation* between the two variables x and y should be computed. The coefficient of correlation is a measure of the degree of linear association between the two variables. It should be pointed out that the value of the coefficient of correlation can be determined for any two variables without specifying one as being depen-

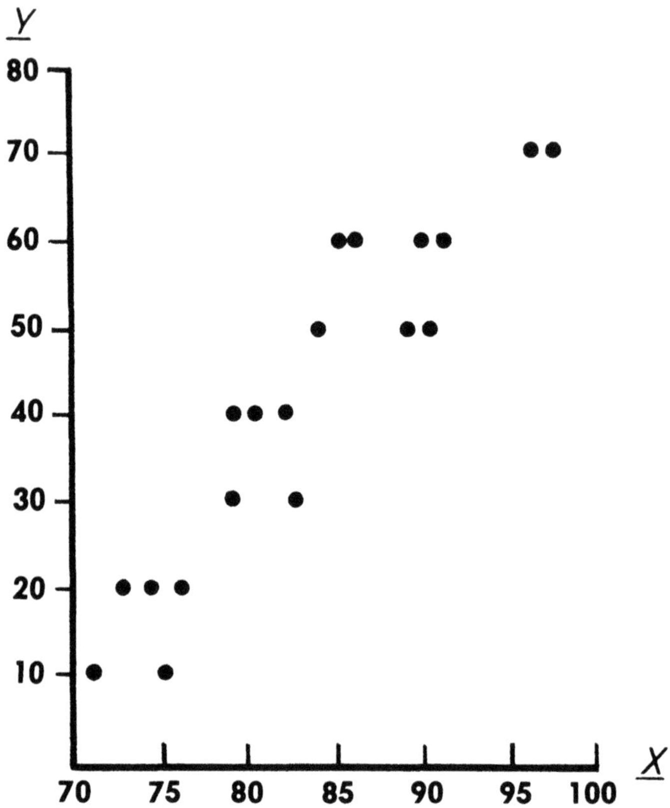

Figure 13.1
Scatter Diagram of Test Scores versus Sales Performance

dent and the other as independent. Test scores were specified as independent in this example because there is an implicit assumption on the part of the firm that sales performance is a function of test scores.

Equation 13.1 defines coefficient of correlation as

$$(13.1) \qquad r = \frac{n\Sigma xy - \Sigma x \Sigma y}{\sqrt{[n\Sigma x^2 - (\Sigma x)^2][n\Sigma y^2 - \Sigma(y)^2]}}$$

Using the test score–sales performance example previously discussed, the coefficient of correlation between x and y can be determined by using Table 13.2 and Equation 13.1.

Table 13.2
Correlation between Employee Test Scores
and Employee Sales Performance, 20 Employees, January 1983

Employee	Test Score (x)	Sales Performance (y)	x^2	y^2	xy
1	75	10	5,625	100	750
2	83	30	6,889	900	2,490
3	78	30	6,084	900	2,340
4	80	40	6,400	1,600	3,200
5	85	60	7,225	3,600	5,100
6	90	50	8,100	2,500	4,500
7	74	20	5,476	400	1,480
8	86	60	7,396	3,600	5,160
9	88	50	7,744	2,500	4,400
10	76	20	5,776	400	1,520
11	96	70	9,216	4,900	6,720
12	82	40	6,724	1,600	3,280
13	71	10	5,041	100	710
14	88	50	7,744	2,500	4,400
15	78	40	6,084	1,600	3,120
16	84	50	7,056	2,500	4,200
17	89	60	7,921	3,600	5,340
18	95	70	9,025	4,900	6,650
19	91	60	8,281	3,600	5,460
20	73	20	5,329	400	1,460
Total	1,662	840	139,136	42,200	72,280

Thus, the coefficient of correlation is

$$r = \frac{20(72{,}280) - (1{,}662)(840)}{\sqrt{[20(139{,}136) - (1{,}662)^2][20(42{,}200) - (840)^2]}}$$

$$= \frac{1{,}445{,}600 - 1{,}396{,}080}{\sqrt{(2{,}782{,}720 - 2{,}762{,}244)(844{,}000 - 705{,}600)}}$$

$$= \frac{49{,}520}{\sqrt{(20{,}476)(138{,}400)}}$$

$$= \frac{49{,}520}{53{,}234} = .93$$

Cause and Effect

The high degree of positive correlation between test scores and sales performance does not necessarily mean that there is a cause-and-effect relationship between the two variables. A high r value can occur by chance alone or because both variables are related to some other variable which has not been considered in the analysis. For instance, there has been an analysis showing a strong positive correlation between teachers' salary increases and sales of liquor. To conclude that changes in the sales of liquor are caused by increases in teachers' salaries is fallacious. In this instance, fluctuations in both variables are no doubt related to changes in a third variable—general economic conditions.

Testing the Significance of the Correlation Coefficient

Since r is a sample correlation coefficient, it is used to make inferences about ρ, the unknown population correlation coefficient. Several statistical tests have been devised to determine whether a correlation coefficient is significant. Only one test of significance will be considered here. The most common inference about ρ is a test of the hypothesis that ρ is actually equal to zero, i.e., that there is no relationship between the two variables. Any relationship between the two variables is purely chance. The null and alternate hypotheses to be tested are

$$H_0: \quad \rho = 0$$
$$H_1: \quad \rho \neq 0$$

The null hypothesis, therefore, states there is no correlation between test scores and sales performance. Stated in reverse, test scores and sales performance are not associated. The alternate hypothesis states that there is a correlation or association between test scores and sales performance.

To test the hypothesis, the t statistic is used. The t value is computed as follows:

$$(13.2) \qquad t = \frac{r\sqrt{n-2}}{\sqrt{1-r^2}}$$

$$= \frac{.93\sqrt{20-2}}{\sqrt{1-.8649}}$$

$$= \frac{(.93)(4.2426)}{.36756}$$

$$= 10.73$$

Assuming that $\alpha = .05$ and with $n - 2$ or 18 degrees of freedom, the critical or tabular value of t is 2.101. Since the computed value of $t = 10.73$, which is greater than 2.101, the null hypothesis that $\rho = 0$ is rejected. Therefore, it can be assumed that a relationship does exist between the two variables. Thus, the statistician retained by the defendant's law firm would be in a position to refute the plaintiff's argument. Based on his sample, test scores appear to be an indication of sales performance.

SIMPLE LINEAR REGRESSION ANALYSIS

Up to this point, data on two variables, x and y, indicate that a linear relationship exists between them. This relationship has been indicated by plotting the data on a scatter diagram, computing a sample coefficient of correlation, and testing the hypothesis about the population coefficient of correlation. A linear relationship can be expressed in the form $y = \alpha + \beta x$, where y represents a value of the dependent variable, x is the independent variable, and α and β are the parameters or coefficients of the regression line. The y intercept of the line is α; it is the value of y when $x = 0$. The coefficient β is simply the slope of the line; it represents the change in y per unit change of x. A typical regression line is illustrated in Figure 13.2. The slope of the line shown in this figure is the ratio of $y_1 - y_0$ to $x_1 - x_0$, i.e.,

$$\beta = \frac{y_1 - y_0}{x_1 - x_0}$$

In most instances the regression relationship is not exact, i.e., for a given value of x, the same value of y will not always result. Instead of an exact relationship between x and y, there is a distribution of y values for each value of x. In that case, the regression line, $\alpha + \beta x$, represents an average value of y for the given x.

In most applications of regression analysis, it is impossible or impractical to obtain all the observations of the population. Consequently, the analysis must be

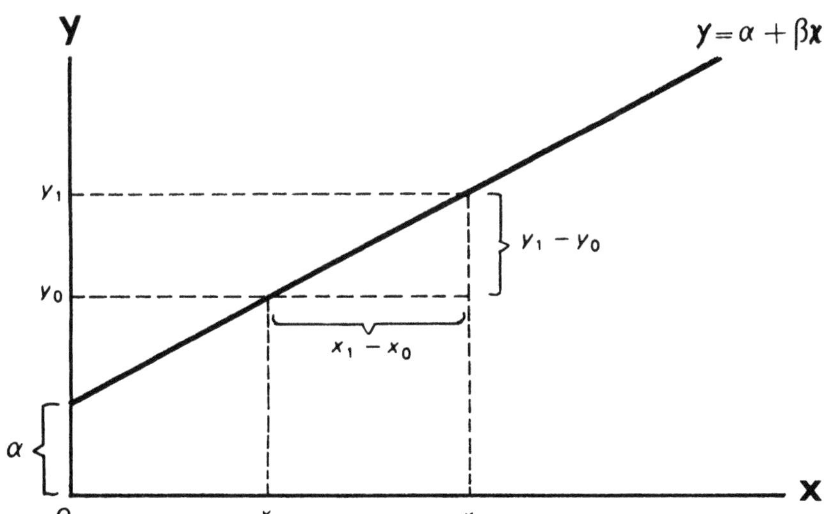

Figure 13.2
Regression Line of y on x

based on sample data. In estimating the regression equation from sample data, the functional relationship is written as

(13.3) $y_c = a + bx$

where

y_c = regression value
a = y intercept
b = slope of regression equation

Equation 13.3 is called the *estimated regression of* y *on* x. The variable y_c is used to designate the computed regression values from the regression equation to distinguish them from the actual y values.

The Least Squares Regression Line

In viewing the scatter diagram of test scores versus sales performance in Figure 13.1, one might be at a loss as to where to draw a straight line that best fits the points. This illustrates the basic problem encountered in selecting a method of fitting a regression line. If the points to which the line is being fitted will not fall on a straight line, it is impossible to find any line that fits perfectly. Thus the problem is to define what constitutes the criterion of *best fit*. Some of the points on the scatter diagram will be above the regression line and others will be below.

The criterion used in regression analysis is called *least squares*. The total deviation of the points on the scatter diagram from the regression line is the sum of all the vertical distances between the points and the line. Least squares analysis assures that the sum of the squared deviations of the plotted points from the regression line is less than for any other line. Least squares analysis has the additional property that the sum of the positive and negative deviations of the observed points about the regression line is 0. These two properties are summarized as follows.

$\Sigma(y - y_c)^2$ is a minimum

$\Sigma(y - y_c) = 0$

The problem of fitting a regression line by the method of least squares requires calculation of the values of a and b which satisfy the least squares criterion. These values are calculated as

(13.4) $a = \bar{y} - b\bar{x}$

and

(13.5) $$b = \frac{\Sigma xy - n\bar{x}\bar{y}}{\Sigma x^2 - n(\bar{x}^2)}$$

Given the data in Table 13.2, the values of $\bar{x}$ and $\bar{y}$ are found to be 83.1 and 42, respectively. By substituting the appropriate values in Equations 13.4 and 13.5, b and a are obtained as follows:

$$b = \frac{72{,}280 - (20)(83.1)(42)}{139{,}136 - (20)(83.1^2)}$$

$$= \frac{72{,}280 - 69{,}804}{139{,}136 - 138{,}112.2}$$

$$= \frac{2{,}476}{1{,}023.8}$$

$$= 2.42$$

$$a = 42 - (2.42)\,(83.1) = -158.97$$

The regression equation (13.3) of y and x then is

$$y_c = -158.97 + 2.42\,b$$

In terms of the original problem, y_c represents the predicted sales performance for a specified value of x test score. Assuming a test score of 75, what is the predicted sales performance of the prospective employee in terms of percentage of the time that he will exceed his weekly sales goal? This can be calculated from the regression equation:

$$y_c = -158.97 + 2.42(75) = 22.53$$

Based on the regression equation, one would interpret to predicted value as follows: Given a test score of 75, it is estimated that the prospective employee will exceed his sales goal 22.53 percent of the time. In a like manner, a prospective employee scoring 90 on the test would be expected to exceed his sales goal 58.8 percent of the time.

In making estimates from a regression equation, it is important to remember that the assumption is being made that the relationship has not changed since the regression equation was computed. Another principle to keep in mind is that the relationship indicated by the scatter diagram may not hold true if the equation is extended beyond the values used in computing the equation.

Standard Error of the Estimate

As was noted earlier, any estimate based on data from a sample is subject to error due to the variation in the population sampled. If a regression equation is used to make estimates, it is desirable to have a measure of the accuracy of these estimates. The more closely the plotted points are grouped around the regression line, the more accurate the estimates based on the regression line are likely to be.

Likewise, the more the points scatter, the less reliable an estimate based on the regression line is likely to be. What is needed is a measure of the degree of scatter of individual items about the regression line.

In simple regression analysis, populations of two variables, x and y, are involved. The measure of the dispersion or scatter of the individual y values around the regression line, y_c, is called the *standard error of the estimate.* Very similar to the standard deviation, it measures the deviations of the individual values of y around the regression line; in other words, the standard error of the estimate is a measure of how much the regression equation varies from the actual data points of y. The sample standard error of the estimate is defined as

$$(13.6) \qquad s_{yx} = \sqrt{\frac{\Sigma(y - y_c)^2}{n - 2}}$$

The degrees of freedom for s_{yx} are $n - 2$; this is because the sample regression line requires estimates of two parameters, α and β. Thus, two degrees of freedom are subtracted from n.

A more convenient computational form for Equation 13.6 is

$$(13.7) \qquad s_{yx} = \sqrt{\frac{\Sigma y^2 - a\Sigma y - b\Sigma xy}{n - 2}}$$

Using the earlier example relating to test scores versus sales performance, the standard error of the estimate of the regression equation $y_c = -158.97 + 2.42\ b$ is

$$s_{yx} = \sqrt{\frac{42{,}200 - (-158.97)(840) - (2.42)72{,}280}{18}}$$

$$= \sqrt{\frac{42{,}200 + 133{,}534.8 - 174{,}917.6}{18}}$$

$$= 6.74$$

Since the standard error of the estimate is an average, sometimes the error in the individual estimates would be more than 6.74 and sometimes less if all possible estimates were made from the regression equation. Assuming the standard errors are normally distributed, about 68 percent of the errors in estimates based on the regression equation will be less than the standard error of the estimate. Thus, the standard error of the estimate enables one to make a probability statement concerning the reliability of estimates from the regression equation. It may be said that the probability is .68 that an estimate made from the regression equation will not deviate more than 6.74 from the true value.

Coefficient of Determination

When using regression analysis, it is often desirable to use the regression relationship to explain the variation of the dependent variable. For example, how much of the variation in the dependent variable is explained or accounted for by its relationship with the independent variable in the regression equation? That is, how much of the variation in y is explained by x?

These questions can be answered by calculating the *coefficient of determination,* which is defined as:

$$(13.8) \quad r^2 = \frac{SSR}{SST}$$

where

SSR = sum of squares explained by the regression line
SST = total sum of squares of y about the regression line

The total sum of squares of y about the regression line can be calculated as follows:

$$(13.9) \quad SST = \Sigma y^2 - n(\bar{y}^2)$$

The amount of variation explained by x, SSR, is defined as

$$(13.10) \quad SSR = b\left[\Sigma xy - \frac{\Sigma x \Sigma y}{n}\right]$$

Therefore, $SST - SSR$, or SSD, represents the total amount of variation in y unexplained by x. This is called the *residual.*

Using the employee test score example, the coefficient of determination between test scores and sales performance can be calculated by first computing SST and SSR (Equations 13.9 and 13.10).

$$SST = 42{,}200 - 20(42^2)$$
$$= 6{,}920$$
$$SSR = 2.42\left[72{,}280 - \frac{(1{,}662)(840)}{20}\right]$$
$$= 5{,}991.9$$

Then, using Equation 13.8:

$$r^2 = \frac{5{,}991.9}{6{,}920} = 86.6$$

The value of r^2, the square of the correlation coefficient, indicates 86.6 percent of the variation in sales performance by employees can be explained by test scores. The word *explained* should be used with caution, since it does not mean that the variation in sales performance was caused by the variation in test scores but that the fluctuations are related to fluctuations in test scores. This means that the remaining 13.4 percent of the fluctuations in sales performance is related to factors other than test scores.

Significance of the Regression

The significance of the regression equation in explaining the variation in the dependent variable can be tested using analysis of variance and the F ratio. This is accomplished by comparing the variation explained by the regression line with the unexplained variation:

$$F = \frac{MSR}{MSD} \tag{13.11}$$

where

MSR = mean square regression
MSD = mean square residual or deviation

The coefficient of determination has been defined as the percentage of the total variance that is explained by the regression equation. The difference between the explained variance and the unexplained variance, expressed as the ratio between two variances, may be tested for a significant difference by use of the analysis of variance.

The three sums of squares—*SSR*, *SSD*, and *SST*—make it possible to set up a table for testing significance. The degrees of freedom associated with the total sum of squared deviations about the regression line are 1 less than the number of individual items in the sample or, in the example used in this chapter, 19. These degrees of freedom are divided into 18 for the unexplained variance and 1 for the variance explained by the regression line. Since there are two constants in the regression equation, 2 degrees of freedom were lost fitting the regression line to the actual values. This means that the unexplained variation uses 18 degrees of freedom, leaving 1 for the explained variation—which corresponds to the one independent variable in the equation. The analysis of variance is illustrated in Table 13.3.

Table 13.3
Analysis of Variance for the Test Score–Sales Performance Regression Model

Type of Variation	Sum of Squared Deviation	Degrees of Freedom	Mean Squares
Explained by regression	5991.9	1	5991.9
Unexplained	928.1	18	51.6
Total	6920.0	19	

The F ratio is the ratio between the variance explained by the regression and the variance that is not explained (Equation 13.11):

$$F = \frac{5{,}991.9}{51.6} = 116.12$$

In testing for significance of the regression, the null hypothesis is one of no relationship between the dependent and independent variable. That is,

H_0: $\beta = 0$
H_1: $\beta \neq 0$

The critical value of F is found in Appendix Table A.5. Since the F value at the .01 level is 8.28, the null hypothesis is rejected. The F value of 116.12 is found to be significant since it far exceeds the tabular value of F with 1 and 18 degrees of freedom. Thus, it is concluded that the regression equation is significant at the .01 level; the independent variable significantly contributes to explaining the variation in the dependent variable.

MULTIPLE REGRESSION

The term *multiple regression* is used to refer to models containing two or more independent variables. The general form of a multiple regression equation is as follows:

$$(13.12) \quad y_c = a + b_1x_1 + b_2x_2 + \ldots + b_nx_n$$

where

y_c = dependent variable
x_1 = independent variable 1
x_2 = independent variable 2
x_n = independent variable n
b_1 = regression coefficient of variable 1
b_2 = regression coefficient of variable 2
b_n = regression coefficient of variable n

Most of the concepts of multiple regression are extensions of simple regression. In real world situations, it often becomes necessary to identify several factors influencing a dependent variable. In economics, for example, the demand for a product—expressed in quantity of the product purchased—involves several factors: price of the product, population, tastes and preferences of consumers, prices of related products, and income of consumers.

In estimating the demand relationship, the influence of each variable must be considered in relation to quantity. In the regression sense, this means that coefficients must be estimated for each independent variable. Since conventional algebraic techniques for multiple regression are cumbersome and extremely time-consuming, solutions are almost exclusively obtained through computer analysis using matrix methods.

To illustrate multiple regression, assume one wants to be able to predict the demand for beef products in this country. The demand function for beef is specified in the following equation:

$$y_c = a + b_1x_1 + b_2x_2 + b_3x_3 + b_4x_4 + b_5x_5$$

where

y_c = total quantity of beef consumed
x_1 = price of beef per pound
x_2 = price of pork per pound
x_3 = total U.S. personal income
x_4 = total U.S. population
x_5 = linear time trend

Data used in the analysis are shown in Table 13.4.

The model resulting from this data is $y_c = -18{,}364.4 - 17{,}639.73x_1 +$

Table 13.4
Beef Consumption and Prices, Pork Prices, Personal Income, and Population, United States, 1964–80

Year	Beef Consumption (millions of pounds)	Beef Price (dollars per pound)	Pork Price (dollars per pound)	Personal Income (billions of dollars)	Population (thousands)
1964	15,399	.451	.472	399.7	180,671
1965	15,890	.427	.471	415.0	183,691
1966	15,867	.464	.491	440.7	186,538
1967	16,896	.417	.464	463.1	189,242
1968	18,965	.398	.458	495.7	191,889
1969	19,261	.433	.542	537.0	194,303
1970	20,355	.442	.587	584.9	196,560
1971	20,730	.451	.544	626.6	198,712
1972	21,358	.473	.537	685.2	200,706
1973	21,600	.492	.580	745.8	202,677
1974	22,030	.490	.542	801.3	204,878
1975	22,213	.547	.534	859.1	207,053
1976	22,647	.576	.626	942.5	208,846
1977	21,413	.696	.810	1,052.4	210,410
1978	23,286	.691	.678	1,115.0	211,901
1979	24,500	.754	.882	1,253.4	213,559
1980	26,480	.644	.855	1,382.7	215,142

Table 13.5
Analysis of Variance for the Beef Demand Regression Model

Type of Variation	Sum of Squared Deviations	Degrees of Freedom	Mean Squares
Explained by regression	154,481,625.98	4	38,620,406.49
Unexplained	3,387,310.96	12	282,275,91
Total	157,868,936.94	16	

$2825.28x_2 + 7.27x_3 + 0.20x_4$. The R^2 value is .97854, meaning that about 97.85 percent of the variation in consumption of beef is explained by the regression equation. The F ratio is 136.82, which is significant at the .01 level. Thus, the independent variables contributed significantly to explaining the variation in the demand for beef products.

Data from which R^2 and F were calculated appear in Table 13.5. Remember that $R^2 = SSR/SST = 154{,}481{,}625.98/157{,}868{,}936.94 = .9785$. The F ratio is $MSR/MSD = 38{,}620{,}406.49/282{,}275.91 = 136.82$.

Note that the degrees of freedom for the total sum of squares is the sample size, n, minus 1, or 16. Degrees of freedom associated with the variation explained by, or attributable to, the regression model are equal to the number of independent variables, or 4.

ILLUSTRATIVE USES OF CORRELATION AND REGRESSION

Several cases have involved the use of correlation and regression. One which illustrated the court's thorough grasp of the concept of correlation, *Boston Chapter, NAACP, Inc. v. Beecher,*[1] has been mentioned previously. In an attempt to validate the use of entrance examinations for prospective firefighters by showing job relatedness of the examinations, the defendant employed a statistician as an expert witness. As the court noted:

> Dr. Costa conducted what he described as a concurrent criteria-related validity study. A brief discussion of terms as described to the Court is in order. The mathematical relationship between the test scores and the measure of job performance is referred to as the correlation coefficient. A correlation coefficient of .00 means the study shows no relationship between the test and job performance, while a coefficient of 1.00 indicates a perfect relationship between the two. In order for a correlation coefficient to have significance, it must be both statistically and practically significant. Statistical significance means that the possibility of the results being reached by chance are minimal. Practical significance means that the coefficient shows a sufficiently high relationship between success on the test and successful job performance. Both Dr. Costa and plaintiffs' expert, Dr. Hunt, agreed that as a ''rule of thumb'' a coefficient of .3 would be the minimum level to indicate a satisfactory relationship.[2]

Dr. Costa used two groups of subjects in his study, which used two performance appraisal systems: system I and system II. System II did not reveal any significant correlations between the two test criteria used, written exam score and general average work, and successful job performance in either group tested. Dr. Costa "admitted that if the results of system II were all that he had available upon which to base a conclusion, he would conclude the exam was not a valid predictor of successful job performance."[3]

Even though Dr. Costa concluded that system I revealed seven significant correlations, the court disagreed:

At any rate, the Court does observe that the Air Mask Operation (.346) and the Loop Man Position (.313) are the only two job tasks which show a significant relationship with the written exam score, and Dr. Hunt described these correlations as being barely significant. . . . Experts have interpreted the EEOC Guidelines to require that at least one relevant criterion be both statistically and practically significant. Dr. Hunt testified that if one significant correlation is to be accepted as evidence that an exam is job related, the one relevant criterion must represent an adequate measure of total job performance factors. If the one criterion only represents a small percentage of the important factors of the job, this should not be considered adequate to support a conclusion of test validity. Although she agreed that the 13 criteria implemented were important, it was Dr. Hunt's opinion that there were only two significant correlations and that those two did not represent an adequate measure of the total factors involved in the job. Based on Dr. Costa's study, she testified that she would not conclude the exam was job related. Dr. Costa is of the opinion that the exam has been shown to be valid according to EEOC Guideline minimum requirements and he has concluded that the exam is a valid predictor of job performance.[4]

Apparently, the court had some difficulty in determining which statistician to believe, finally concluding that the firefighters exam was not related to job performance:

The Court has had no little difficulty in examining the testimony of both experts and in interpreting Dr. Costa's study and the results derived therefrom. Nor has the Court relished the thought of determining which expert opinion to accept on a subject about which the Court knows very little. It is easy for a member of the legal profession, however, to understand how two competent professionals can come to different conclusions given the same set of facts. And it appears, as with the law, that the area of psychological testing is not one where elements of preciseness and exactness usually obtain. . . . The Court has concluded that defendants have not met their burden of demonstrating that the exam is "in fact substantially related to job performance."[5]

Another case in which regression and correlation played a significant role is *Albermarle Paper Company v. Moody,*[6] a Title VII case heard by the Supreme Court. One portion of the case dealt with use of employment tests and their validation for job relatedness. The court held that such tests are impermissible unless shown, by professionally acceptable methods, to be "predictive of or significantly correlated with important elements of work behavior which comprise or are related to the job or jobs for which candidates are being evaluated.''[7]

Albermarle used two general ability tests—the Beta Examination for nonverbal intelligence and the Wonderlic Test (forms A and B), a purported measure of general verbal facility. Applicants for hire into various skilled lines at the plant were required to score 100 on the Beta Examination and 18 on one of the two forms of the Wonderlic Test. Albermarle ''added the Wonderlic Tests in 1963, for the skilled lines, on the theory that a certain verbal intelligence was called for by the increasing sophistication of the plant's operations.''[8] The Court noted, however, that the company ''made no attempt to validate the test for job relatedness, and simply adopted the national norm of 18 as a cut-off point for new job applicants.''[9]

Addressing the question of test validation the Court noted:

> For each job grouping, the expert computed the ''Phi coefficient'' of statistical correlation between the test scores and an average of the two supervisorial rankings. Consonant with professional conventions, the expert regarded as ''statistically significant'' any correlation that could have occurred by chance only five times, or fewer, in 100 trials. On the basis of these results, the District Court found that ''[t]he personnel tests administered at the plant have undergone validation studies and have proven to be job related.'' Like the Court of Appeals, we are constrained to disagree. . . . Albermarle's validation study is materially defective in several respects: Even if it had been otherwise adequate, the study would not have ''validated'' the Beta and Wonderlic test battery. . . . The study showed significant correlations for the Beta Exam in only three of eight lines. Though the Wonderlic Test's Form A and Form B are in theory identical and interchangeable measures of verbal facility, significant correlations for one form but not for the other were obtained in four job groupings. In two job groupings neither form showed a significant correlation.[10]

As noted in Chapter 1, multiple regression has proved especially useful in legal proceedings. In addition to use by the regulatory agencies cited in Chapter 1, the technique has been used in employment discrimination cases to isolate the effect of race or sex. In *Smith v. Reader's Digest Association,*[11] multiple regression analysis indicated significant discrimination against women after controlling for variables related to age, experience, and education. Given this evidence, the defendant agreed to an award of back pay.

NOTES

1. 371 F. Supp. 507 (1974).
2. *Id* at 516.
3. *Id* at 516–17.
4. *Id* at 517.
5. *Id.*
6. 422 U.S. 405 (1975).
7. *Id* at 406.

8. *Id* at 428.
9. *Id*.
10. *Id* at 430–31.
11. 73 Civ. 4883 (S.D.N.Y., filed Nov. 14, 1973).

SELECTED REFERENCES

Berkson, Joseph. "Smoking and Lung Cancer." *The American Statistician*, Vol. 17 (October 1963), pp. 15–22.

Curtis, Wayne C. "Predicting Sectoral Employment in Counties and Multi-County Areas: A Note." *Review of Business and Economic Research*, Vol. 12 (Fall 1976), pp. 73–78.

Draper, Norman D. and Harry Smith. *Applied Regression Analysis*. New York: John Wiley & Sons, 1966.

Duchan, Alan I. "A Relationship Between the *F* and *t* Statistics and the Simple Correlation Coefficients in Classical Least Squares Regression." *The American Statistician*, Vol. 23 (June 1969), pp. 27–28.

Fisher, Franklin M. "Multiple Regression in Legal Proceedings." *Columbia Law Review*, Vol. 80 (May 1980), pp. 702–36.

Fisher, R. A. *Statistical Methods for Research Workers*, Chaps. 5–7. 14th ed. New York: Hafner Publishing Co., 1969.

Geary, R. C. and C.E.V. Lesser, "Significance Tests in Multiple Regression." *The American Statistician*, Vol. 22 (February 1968), pp. 20–21.

Harder, T. and F. U. Pappi. "Multiple-Level Regression Analysis of Survey and Ecological Data." *Social Science Information*, Vol. 7, No. 5 (July 1960), pp. 48–67.

Suits, Daniel B. "The Use of Dummy Variables in Regression Analysis." *Journal of the American Statistical Association*, Vol. 52 (December 1959), pp. 548–51.

CHAPTER 14

INDEX NUMBERS

Index numbers are special ratios whose purpose usually is to show relative change over time—how much something has changed from one time period to another. Index numbers associated with business and economics are often described as barometers of economic activities. Variations in the indexes of sales, prices, production, new orders, and many other items are closely watched. Most attorneys, particularly those involved in arbitration or wage disputes, are commonly exposed to such index numbers as the consumer price index (CPI) and the producer price index (PPI), formerly called the wholesale price index (WPI). The CPI is widely used in so-called escalator clauses in collective bargaining; these clauses provide for adjustments of wages according to the increase or decrease in the CPI. In many instances, the CPI has been applied to adjust alimony and child support payments.

In its simplest form, an index number is the ratio of two numbers expressed as a percent. It is a *percentage relative* that compares measures in a given time period with the same measures at a fixed time period in the past. When changes in only one variable occur, such as the total output of steel produced in the United States each year, it is a simple matter to express the amount produced each year as a percent of the number produced in some given base year. To compute a cost of living index, on the other hand, is a complex matter which may involve a number of technical questions as well as being costly. Such an index must consider all the important goods and services purchased by thousands of consumers in hundreds of locations. In both examples, however, the purpose of the index is to give a quick snapshot of changes occurring over time.

This chapter should be of particular interest to many practicing attorneys. Its purpose is to summarize briefly the basic techniques and theory of index numbers and to examine several of the more important indexes produced by governmental and private agencies to see how these indexes are put together and what they measure.

KINDS OF INDEX NUMBERS

The usefulness of index numbers is determined by the types of questions they can answer. Each index number is designed for a particular purpose; it is this specific purpose that determines its method of construction. Two basic types of index numbers will be considered in this chapter: price indexes and quantity indexes. Emphasis will be placed on the former.

Indexes of prices include such well-known indexes as the consumer price index, the producer price index, and the Dow-Jones stock averages. All of these indexes are designed to measure changes in prices while holding measures of quantity constant. A *quantity index,* on the other hand, measures changes in the volume of goods produced, bought, or consumed. There are numerous quantity indexes computed, but perhaps the most broadly used one currently published in the United States is the index of industrial production.

CONSTRUCTION OF INDEX NUMBERS OF PRICES

Price index numbers are probably the most common and useful indexes. Therefore, they will be used to provide general illustrations of index number construction. Other types of indexes are constructed using the same general procedures. Since there are literally dozens of ways to construct a price index, the approach here is to begin with the simplest procedures and proceed to the more complicated methods.

Simple Price Relatives

As noted earlier, a relative is simply the ratio of two numbers expressed as a percent. A *price relative* is the ratio of two prices, a current price, designated p_i, and a base period price, p_o. To construct a simple index number, i.e., one involving the price of a single commodity, the index for period i is

$$(14.1) \quad P_i = \frac{p_i}{p_o}(100)$$

where

P_i = price index for period i
p_i = price of commodity in period i
p_o = price in the base period

To illustrate, assume a law firm wants to construct a price index for the average hourly wage rate paid to secretarial workers in 1982 using 1975 as the base period. In other words, the firm wishes to know how the average secretarial wage rate in 1982 compares with that in 1975 on a *relative* basis. Assuming that average secretarial wage was $3.15 in 1975 and $4.35 in 1982, the 1982 index might be constructed as follows:

$$P_i = \frac{4.35}{3.15}(100) = 130.10$$

The price index for 1982 indicates that the average hourly wage paid in 1982 was 130.1 percent of that paid in 1975.

The technique illustrated in Equation 14.1 is limited in that it can be used when only one commodity or class of items is being considered. When more than one series or class is involved, an average of price relatives must be considered.

Unweighted Average of Price Relatives

An unweighted average of price relatives may be used when changes in the price of a number of items or commodities are desired. The assumption is made that all the series being considered are of equal importance or weight. This index is constructed by first converting each reported price for each item or commodity into a simple relative by dividing by the base period price of that commodity or item. These relatives are then summed for each year, the sum is multiplied by 100, and then it is divided by the number of commodities. By converting all prices to relatives and finding their averages, the effect of the measurement unit is eliminated. The formula for the unweighted average of relatives index is

$$(14.2) \qquad P_i = \frac{\Sigma p_i/p_o(100)}{n}$$

where n represents the number of items entering into the index.

Assume the law firm mentioned earlier wishes to construct an aggregate index of unit prices paid for certain selected expense items for 1982 using 1975 as the base period. The expense items are listed below:

Expense Item	*Measurement Unit*	*1975*	*1982*	p_i/p_o
Secretarial wages	hour	$ 3.15	$ 4.35	1.381
Rent	square foot	5.00	12.00	2.400
Automobile lease	each	3,480.00	5,400.00	1.552
Custodial wages	hour	2.50	3.60	1.440
Photocopying	page	.05	.12	2.400
$\Sigma p_i/p_o$				9.173

Using Equation 14.2, the price index, P_i, is calculated as follows:

$$P_i = \frac{9.173 \times 100}{5} = 183.5$$

Thus, the unweighted average of the price relatives index for the selected cost items in 1982 is 183.5. This means that the average index, assuming equal weights for all cost components, is 183.5; this index number indicates that the prices of the selected items increased by 83.5 percent from 1975 to 1982. As will be noted from the table above, price relatives for individual items vary from the average of all relatives.

Weighted Average of Price Relatives

The assumption made in the previous section—that all the items or series being averaged are of equal importance—can seldom be made in actual practice. Some of the series are almost always more important than others, and their relative positions need to be recognized with an appropriate system of weighting when their relative price changes are combined into an index. In determining a system of weights, the statistician wants to be sure that it accurately represents the relative importance of each series to be included in the index. Further, the weights must be units that can be added.

The most common type of weight used in constructing a price index is quantity. This assumes that the quantity of a commodity sold or purchased in a given time period is a valid measure of its comparative importance. In the prior example, the comparative importance of each component of the index should be proportional to the quantity of each item used by the firm. Nevertheless, the question remains: What quantity should be used as weights?

The two most common weighting systems are named for the individuals most influential in their development. When an index number is constructed using *base year* quantities as weights, it is known as *Laspeyres' price index.* The formula for the Laspeyres' weighted price index is

$$(14.3) \qquad P_i = \frac{\Sigma p_i q_o}{\Sigma p_o q_o} (100)$$

This index measures the change in the total cost of the selected items used by the law firm. Exactly how long this pattern of expenditures should be used as a weighting factor is problematical. Expenditure patterns change with time, and when a pattern changes, the base period should also be changed.

Another well-known but less used price index is *Paasche's price index.* This index uses *current year* quantities as weights. The formula for a Paasche weighted index is

$$(14.4) \qquad P_i = \frac{\Sigma p_i q_i}{\Sigma p_o q_i} (100)$$

The Paasche index measures the change in total cost of the selected items that represent an expenditure pattern typical of the current year. Use of current year weights avoids the problem of changing patterns. However, it requires that the weighting system be continually reestablished. Because of the difficulty involved in establishing a new set of weights and applying them each period, the Laspeyres method is by far the more commonly used.

Suppose the law firm decides that the unweighted price index numbers developed earlier are not truly indicative of the actual price increases because of defects in that index. Partners in the firm decide to construct a new index for 1982 relative to the base year, 1975. The Laspeyres index is presented in Table 14.1.

Using Equation 14.3, the weighted index number for 1982 is

$$P_i = \frac{211{,}452}{126{,}148}(100) = 167.6$$

This index number for 1982 indicates that the cost of the selected items increased by 67.6 percent between 1975 and 1982. This compares with the unweighted index number of 183.46.

Chain Index Numbers

The index number formulas discussed previously use a fixed base period. In each case, the price index for a given period made it possible to compare prices in that period with prices in some fixed period. It is sometimes more convenient and practical to compute an index that has as its base the time period just past—the most recent time period prior to the current year. This is particularly true with such indexes as the consumer price index or producer price index, both of which are published by the Bureau of Labor Statistics of the U.S. Department of Labor. In constructing these indexes, the samples will vary from one month to another; this is because the products being priced may change over time. It is much easier to obtain information that is comparable for two adjoining time periods than it is to get comparable data over longer periods of time. A *chain index* is one that has a moving base—the period immediately preceding the period of the index.

Table 14.1
Laspeyres' Price Indexes for Five Cost Items, 1975–82 (1975 = 100)

Item	Quantity Purchased, 1975	Price 1975 (dollars)	Price 1982 (dollars)	$p_i q_o$	$p_o q_o$
Secretarial wages	13,520 hours	3.15	4.35	58,812	42,588
Rent	5,500 square feet	5.00	12.00	66,000	27,500
Automobile lease	12 each	3,480.00	5,400.00	64,800	41,760
Custodial wages	5,200 hours	2.50	3.60	18,720	13,000
Photocopying	26,000 pages	.05	.12	3,120	1,300
				211,452	126,148

The formula for a chain price index for a current period, i, based on prices in the previous period, $i - 1$, may be written

$$(14.5) \qquad P_i = \frac{\Sigma p_i q_o}{\Sigma(P_i - 1)(q_o)}$$

where

p_i = price in period i
p_{i-1} = price in a period immediately preceding period p_i
q_o = quantity in fixed base period

As the astute reader will note, this formula is a modified version of the Laspeyres' index.

SOME COMMON INDEX NUMBERS

Attorneys may frequently be required to use index numbers constructed and reported by someone else. Many of the series of index numbers useful to attorneys are developed and reported by agencies of the U.S. government. Discussed in this section are three of the most important indexes.

Consumer Price Index

The consumer price index (CPI) is a monthly statistical measure of the average change in prices in a fixed market basket of goods and services. Effective with the January 1978 index, the Bureau of Labor Statistics of the U.S. Department of Labor began publishing CPIs for two groups of the population. One index, a new CPI for all urban consumers, covers 80 percent of the total noninstitutional population; the other index, a revised CPI for urban wage earners and clerical workers, covers about half the new index population. In addition to wage earners and clerical workers, the all urban consumers index includes professional, managerial, and technical workers, the self-employed, short-term workers, the unemployed, retirees, and others not in the labor force.

The CPI is based on prices of food, clothing, shelter, fuel, drugs, transportation fares, doctors' and dentists' fees, and other goods and services that people buy for day-to-day living. It is designed to measure the effect of price changes on about 400 consumer items. The quantity and quality of these items is kept essentially unchanged between major revisions so that only price changes will be measured. Prices are collected from about 18,000 tenants, 24,000 retail establishments, and 18,000 housing units for property taxes in 85 urban areas across the country. All taxes directly associated with the purchase and use of items are included in the index. Because the CPIs are based on the expenditures of two population groups, they may not accurately reflect the experience of individual families and single persons with different buying patterns.

That this index is important is reflected by the fact that the incomes of about half of the U.S. population are pegged to the CPI. Collective bargaining contracts covering almost 10 million workers tie wages to this index. Pensions and other incomes are adjusted to changes in this index for over 50 million social security beneficiaries, retired military and federal civil service employees and survivors, and other groups.

In addition to the two overall consumer price indexes, individual indexes are also constructed for various subgroups such as food, housing, apparel, and upkeep. Also, indexes are published for most of the larger cities in the United States.

The CPI has been published continuously since 1919. The base period is moved forward every ten years or so, and the goods and services included in the index are revised periodically. The current base period for this index is 1967. A standard chain index formula such as Equation 14.5 is used to calculate the CPI from prices for the market basket items. A sample CPI for selected items is shown in Table 14.2.

Producer Price Index

The producer price index (PPI), also compiled by the Bureau of Labor Statistics, is designed to measure average changes in prices of all commodities sold in the primary markets of the United States. This index shows the rate and direction of changes in the primary market price of approximately 2,800 different commodities. The sample used for calculating these indexes contains about 10,000 quotations per month selected to represent the movement of prices of all commodities produced in the manufacturing, agriculture, forestry, fishing, mining, gas and electricity, and public utilities sectors. The population includes all commodities produced or imported for sale in commercial transactions in primary markets in the United States.

PPIs can be organized by stage of processing or by commodity. The stage of processing structure organizes products by degree of fabrication, i.e., finished goods, intermediate or semifinished goods, and crude materials. The commodity structure organizes products by similarity of end use or material composition.

To the extent possible, prices used in calculating producer price indexes apply to the first significant commercial transaction in the United States from the production or central marketing point. Price data are generally collected monthly, primarily by mail questionnaire. Most prices are obtained directly from producing companies on a voluntary and confidential basis.

The reference base for the index is 1967, i.e., 1967 = 100. The PPI has been published continually since 1890. Table 14.3 provides a sample of the PPI.

Index of Industrial Production

Unlike the indexes measuring changes in prices, the index of industrial production (IIP), compiled and published by the Board of Governors of the Federal Reserve System, measures changes in physical volume of output from the na-

Table 14.2
Consumer Price Index for Selected Items, All Urban Consumers, 1981–82
(1967 = 100)

	1981			1982			
Item	*April*	*November*	*December*	*January*	*February*	*March*	*April*
All items	266.8	280.7	281.5	282.5	283.4	283.1	284.3
Food and beverages	265.7	269.9	270.5	273.6	275.8	275.6	276.5
Housing	284.8	304.2	305.2	306.1	307.8	306.7	309.4
Rent, residential	204.2	215.0	216.5	217.8	218.6	219.6	220.1
Homeownership	339.3	367.2	367.8	367.5	368.7	365.7	370.6
Fuel and other utilities	310.5	329.8	331.8	336.2	337.1	339.3	339.2
Apparel and upkeep	186.4	191.3	190.5	187.3	188.0	191.1	191.9
Transportation	275.3	289.1	289.8	289.9	288.0	285.1	282.9
Medical care	287.0	308.2	310.2	313.4	316.2	318.8	321.7
Entertainment	219.2	226.8	227.3	229.2	231.2	232.8	233.9
Personal care	228.7	237.7	239.1	240.9	242.3	243.7	245.9
All commodities	250.8	258.0	258.4	258.8	259.5	258.8	258.9
All services	295.4	320.6	321.8	323.9	325.3	325.5	328.4

SOURCE: U.S. Department of Labor, Bureau of Labor Statistics, *Monthly Labor Review*, Washington, D.C.: Government Printing Office, July 1982, pp. 81–86.

Table 14.3
Producer Price Indexes for Selected Items, by Stage of Processing, 1981–82
(1967 = 100)

	1981		1982				
Commodity Grouping	*May*	*December*	*January*	*February*	*March*	*April*	*May*
Finished goods	269.9	275.4	277.9	277.4	276.9	276.9	277.7
Finished consumer goods	271.5	275.8	278.3	278.1	277.2	276.9	277.6
Capital equipment	262.5	274.1	276.2	274.8	275.7	277.1	278.3
Intermediate materials	306.7	309.4	311.0	311.3	310.9	310.1	309.8
Materials for manufacturing	285.1	289.3	290.4	291.3	290.8	290.9	291.5
Materials for construction	288.5	291.1	292.0	292.8	293.3	293.8	293.4
Processed fuels	608.7	598.1	604.4	597.1	593.5	579.8	569.9
Containers	276.4	280.2	282.5	285.2	286.5	287.4	287.1
Supplies	264.0	268.3	269.8	270.7	270.9	272.3	273.6
Crude materials	334.4	311.5	318.4	321.5	319.9	322.8	328.1
Foodstuffs and feedstuffs	260.6	233.7	242.6	248.3	247.9	254.3	262.3
Nonfood materials	492.4	478.6	481.5	479.3	475.0	470.4	470.4

SOURCE: U.S. Department of Commerce, Bureau of Labor Statistics, *Monthly Labor Review*, Washington, D.C.: Government Printing Office, July 1982, p. 89.

Table 14.4
Indexes of Industrial Production for Selected Items, 1981–82
(1967 = 100)

	1981			1982					
Grouping	*May*	*June*	*December*	*January*	*February*	*March*	*April*	*May*	*June*
Total index	152.7	152.9	143.4	140.7	142.9	141.7	140.2	139.4	138.4
Consumer goods	150.7	150.3	142.0	139.6	141.8	141.5	142.2	143.3	143.4
Durables	147.3	147.9	123.2	120.1	125.9	128.1	130.7	133.6	135.9
Nondurables	152.1	152.3	149.5	147.4	148.1	146.8	146.8	147.2	146.4
Equipment	152.1	153.0	152.1	147.2	147.3	145.9	143.3	140.9	138.3
Business	182.0	183.6	179.0	172.2	171.6	151.2	145.9	139.8	134.0
Defense	102.0	101.7	107.0	105.2	106.5	125.6	122.9	122.4	121.5
Intermediate products	156.1	154.9	145.9	143.4	146.3	145.2	143.5	142.6	141.6
Construction	146.5	143.4	127.0	124.2	127.5	164.6	164.0	162.7	—
Business	165.6	166.2	164.6	162.4	165.1	184.5	185.0	180.8	—
Materials	153.4	154.0	139.0	137.2	140.4	138.5	136.1	134.8	133.9
Durables	152.8	152.4	134.0	129.7	132.4	130.7	127.8	126.4	125.8
Nondurables	179.0	176.9	158.3	156.8	164.2	162.0	160.3	158.2	156.7
Energy	123.0	129.3	127.4	130.9	130.3	128.2	126.1	125.8	125.0

SOURCE: *Federal Reserve Bulletin*, Washington, D.C.: Board of Governors, Federal Reserve System, July 1982, p. A48.

NOTE: Dashes indicate data were not reported.

tion's factories, mines, and utilities. This index is constructed by combining 235 individual monthly series for manufactured goods and mineral products representing about 80 percent of the nation's output of goods and services. Separate indexes are constructed for various groups and subgroups. It does not include agricultural production, transportation, construction, or the trade and service industries.

First computed in the 1920s, the IIP is the most widely used of all the current indexes as an indicator of general business conditions. This index is published monthly in the *Federal Reserve Bulletin*. Table 14.4 is a sample of the IIP.

PROBLEMS OF INDEX NUMBER CONSTRUCTION

The indexes and formulas discussed in this section provide little indication of the many practical problems involved in their use. The fact that there are many formulas suggests that there is no best one to use; the one selected for the particular index will depend on the particular problems associated with that index.

Attorneys who may be potential users of any of the indexes should be aware of the major problems encountered in the construction of index numbers. There are four major categories of problems, each of which will be discussed in this section.

Items to Be Included

The purpose for which the index is being computed is basic in determining items to be included. An index of medical costs, for example, must be based on a carefully selected set of medical items representative of the intended index group. If the index is constructed from a sample rather than from a complete census of items—a common occurrence—the choice of specific commodities included in the index becomes extremely important. This choice cannot be made by a chance selection process; it must relate to the purpose of the index and the nature of the items represented.

To illustrate the point, the Bureau of Labor Statistics has determined that the typical family—having moderate income and residing in an urban area—purchases about 2,000 items. Of these, only about 400 are included in the CPI. In deciding which of the 2,000 items should be included, consideration had to be given to the relative importance of each item and its effect upon the living costs of the typical family for which the index is constructed.

Quality and Price

Those who construct indexes continually face the problem of measuring price changes without incorporating changes in the quality of items. Great care must be exercised in selecting commodities that remain constant in quality over the period of time covered. To ensure that a price index reflects only changes in price and not changes in quality, detailed specifications are required to describe the items to be priced. In addition, revisions must be made to provide for changes in buying patterns and for new products.

Base Period

Two considerations influence the choice of a base period for an index. First, the base period should be "typical" or "normal" for the activity to be measured by the index. The definition of "typical" is somewhat subjective; in general, a time period should be selected that exhibits relative stability with respect to the variable(s) being observed. Second, the period should be recent enough so that comparisons with the base are meaningful. For example, constructing an index of current prices using 1930 as a base would be of no use in assessing recent price movements.

At present, the year 1967 is used as a base period for most indexes; for the remainder, 1972 is the base. In the past, more than one year has been used as a base. The use of several years as a base provides a smoothing effect on year-to-year fluctuations. For this reason, the periods 1947–49 and 1957–59 were used as base periods. Researchers have found it is now necessary to shift the base period for most index numbers about every ten years to keep pace with the rapid rate changes that are occurring in the economy.

Weighting System

A realistic price index must be a weighted index. Since the only unit that is common to all commodities is value, it is necessary to use weights based on values. That is, in constructing an index, the price of each commodity is generally weighted by the quantity sold or purchased during some period. The two common weighting systems—the Laspeyres and the Paasche systems—were discussed earlier.

ILLUSTRATIVE USES OF INDEX NUMBERS

Index numbers have played a significant role in numerous cases. For example, within recent years indexes have enjoyed a prominent position in establishing reasonable damages in negligence cases involving human injury or death. Models developed by economists and statisticians have sought to provide a definitive quantitative means of establishing the loss of earnings resulting from physical or mental impairment or death. Procedures employed parallel, for the most part, those developed by the U.S. Department of Commerce for estimating lifetime earnings of individuals.

It is generally conceded that changes in the general price level greatly affect lifetime earnings of individuals and will continue to influence earnings in the future. Therefore, some means of accounting for price level changes must be incorporated into the modeling procedure as illustrated below:[1]

$$(14.6) \quad FGP_n = \sum_{i=1}^{n} B_i(1 + \alpha)^{n-1}(1 + \beta)^{n-1}$$

where

FGP_n = expected future value of lifetime earnings prior to impairment or death, with adjustments for changes in productivity and the general price level, after n years
B_i = expected annual earnings at the end of ith year
α = expected annual rate of change in earnings due to productivity influences
β = expected annual rate of change in earnings due to price level influences
n = number of years of remaining worklife
i = year of impairment where $i = 1, 2, \ldots, n$

The β component in the model relies almost exclusively on past and expected future changes in the general price level as reported in the consumer price index. Admissability of estimates involving projections of earnings of the type discussed here is becoming increasingly more common. Rulings in numerous court cases have upheld the validity of this approach.[2]

Another valuable use of index numbers is in the area of collective bargaining, particularly cost-of-living escalator clauses.[3] The major thrust for the use of the escalator clause in collective bargaining came from negotiations between General Motors and the United Automobile Workers shortly after World War II. At that time, the typical collective bargaining agreement was for one year's duration. The UAW and General Motors, however, reached an agreement whose term was five years.

A key factor in the development of that long-term contract was the agreement between the parties on a cost-of-living escalator. It provided for an automatic adjustment in hourly wages every quarter (three months), based upon changes in the consumer price index. Depending on the movement of the CPI, wages could be adjusted both upward or downward. In no event, however, could wages fall below the levels established in the new agreement. Thus, there was a floor but no ceiling.

The escalator clause gave assurance to workers that rising prices would not destroy the standard of living which their newly negotiated wage rates would provide. In so doing, the escalator clause removed a major stumbling block to a multi-year collective bargaining agreement.

The UAW–GM agreement had one other feature that has also become a landmark in collective bargaining. It contained provision for an "annual improvement factor"—that is, an increase (in addition to any escalator adjustments that might occur) based upon the national productivity growth rate. The purpose of this annual improvement factor was to assure that, during the five years, workers would not only keep pace with the increase in inflation, but they would also receive their share of any output increases attributed to productivity changes.

Escalator clauses have become fairly commonplace since that period, even in agreements whose term is less than five years. They are still rather rare in agreements of less than two years' duration. Although they exist predominantly in the private sector, they have begun to spread, in one form or another, into public sector bargaining.

Like any other provision in the collective bargaining agreement, the details of an escalator clause are a matter for negotiations between the parties. Consequently, there are wide variations to be found in the design of escalator clauses from one agreement to the next. But they all have the same basic purpose—to protect the buying power of workers' wages and salaries from erosion due to rising prices. The extent to which any particular clause provides the desired protection depends upon the agreement the parties reach with respect to a number of important elements of the escalator clause.

To indicate the content and structure of an escalator clause, two sample clauses appear below. One provides for percentage adjustments, while the other uses a dollars-for-points approach. The same set of assumptions has been used for both clauses.

Sample Escalator Clause—Straight Percentage

Section 1. In addition to the basic salary scales specified in this Agreement, all employees covered by this Agreement shall be paid a cost-of-living allowance, to be determined on the basis of changes in the Consumer Price Index—All Items, for the United States (1967 = 100), published by the Bureau of Labor Statistics of the U.S. Department of Labor.

Section 2. The cost-of-living allowance shall be based on changes in the Consumer Price Index between May 1982 and each of the following months:

1982	*1983*	*1984*
August	February	February
November	May	May
	August	
	November	

Section 3. The percentage change, if any, between the Consumer Price Index for May 1982 and for each of the months listed above shall be applied to the basic salary for each classification, computed to the nearest dollar, in order to determine the cost-of-living allowance that shall be paid to each employee during each subsequent quarterly (three-month) period.

Section 4. The cost-of-living allowance shall be paid beginning with the start of the first pay period that commences following publication of the Consumer Price Index for each of the months listed above.

Section 5. The cost-of-living allowances shall be used in the computation of straight time, overtime, and all other allowances and benefits.

Section 6. This cost-of-living provision shall be applied to increase or reduce pay, but in no event shall it operate to reduce the salaries below the basic salary scales specified in this Agreement.

Section 7. No adjustments, retroactive or otherwise, shall be made due to any revision which may later be made in the published figures for the Consumer Price Index.

Sample Escalator Clause—Dollars-for-Points

Section 1. In addition to the basic salary scales specified in this Agreement, all employees covered by this Agreement shall be paid a cost-of-living allowance, to be determined on the basis of changes in the Consumer Price Index—All Items, for the United States (1967 = 100), published by the Bureau of Labor Statistics of the U.S. Department of Labor.

Section 2. The cost-of-living allowance shall be based on changes in the Consumer Price Index between May 1982 and each of the following months:

1982	*1983*	*1984*
August	February	February
November	May	May
	August	
	November	

Section 3. The cost-of-living allowance for each employee covered by the Agreement shall be one dollar per week for each 0.62 points change in the Consumer Price Index between May 1982 and each of the months listed above.

Section 4. The cost-of-living allowance shall be paid beginning with the start of the first pay period that commences following publication of the Consumer Price Index for each of the months listed above.

Section 5. The cost-of-living allowance shall be used in the computation of straight time, overtime, and all other allowances and benefits.

Section 6. The cost-of-living provision shall be applied to increase or reduce pay, but in no event shall it operate to reduce the salaries below the basic salary scales specified in this Agreement.

Section 7. No adjustment, retroactive or otherwise, shall be made due to any revision which may later be made in the published figures for the Consumer Price Index.

These sample clauses bring into focus the fundamental distinction between the two approaches in designing the formula for an escalator clause. The total amount of money received by the bargaining unit would be the same in both cases. In one approach—the second sample clause—everyone in the bargaining unit receives identical amounts. In the other—the percentage approach—the same total amount of money is involved, but it is distributed differently among the members of the bargaining unit. In the latter case, the distribution is based on the employee's salary classification.

NOTES

1. For a more detailed discussion of this concept, see Wayne C. Curtis and Lowell E. Wilson, "A Model to Predict Loss of Earnings from Impaired or Destroyed Capacity," *Atlantic Economic Journal,* Vol. 6 (July 1978), pp. 41–46.

2. Perry v. Allegheny Airlines, Inc., 489 F. 2d 1349 (1974); Willmore v. Hertz Corp., 437 F. 2d 357 (1971); Petition of United States Steel Corp., 436 F. 2d 1256 (1970), cert. den. 402 U.S. 987, 91 S. Ct. 1649, 29L. Ed. 2d 153; reh. den. 403 U.S. 924, 91 S. Ct. 2247, 29L. Ed. 270; McCauley v. United States, 470 F. 2d 137 (1972); Beaulieu v. Elliot, 434 p. 2d 665 (1967); Resner v. The Northern Pacific Railroad, 502 p. 86 (1972); Plurod v. The Southern Pacific Transportation Corp., 513 P. 2d 1140 (1973).

3. Adapted from U.S. Department of Commerce, Bureau of Labor Statistics, *The Use of Economic Data in Collective Bargaining,* Washington, D.C.: Government Printing Office, 1978.

SELECTED REFERENCES

Fisher, Irving. *The Making of Index Numbers.* Boston: Houghton Mifflin Co., 1927.

Midgett, Bruce D. *Index Numbers.* New York: John Wiley & Sons, 1951.

Mundlak, Y. and A. Razin. "Aggregation, Index Numbers, and the Measurement of Technical Change." *Review of Economics and Statistics,* Vol. 50 (May 1969), pp. 166–75.

Persons, Warren M. *The Construction of Index Numbers.* Boston: Houghton Mifflin Co., 1928.

U.S. Department of Labor, Bureau of Labor Statistics. *BLS Handbook of Methods.* Bulletin 1910. Washington, D.C.: Government Printing Office, 1976.

———. *Relative Importance of Components in the Consumer Price Indexes.* Report 595. Washington, D.C.: Government Printing Office, 1980.

———. *The Use of Economic Data in Collective Bargaining.* Washington, D.C.: Government Printing Office, 1978.

APPENDIX: STATISTICAL TABLES

Table A.1
Cumulative Binomial Probability Distribution

$$P(r \geq \tilde{r} \mid n, p) = \sum_{r=\tilde{r}}^{n} \binom{n}{r} p^{r} q^{n-r}$$

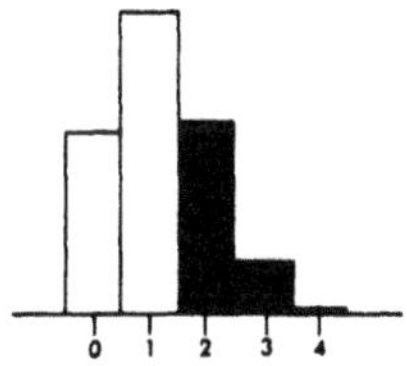

$P(r \geq 2 \mid n = 4, p = 0.30) = 0.3483$

n = 1

r \ p	.01	.02	.03	.04	.05	.06	.07	.08	.09	.10
1	.0100	.0200	.0300	.0400	.0500	.0600	.0700	.0800	.0900	.1000
	.11	.12	.13	.14	.15	.16	.17	.18	.19	.20
1	.1100	.1200	.1300	.1400	.1500	.1600	.1700	.1800	.1900	.2000
	.21	.22	.23	.24	.25	.26	.27	.28	.29	.30
1	.2100	.2200	.2300	.2400	.2500	.2600	.2700	.2800	.2900	.3000
	.31	.32	.33	.34	.35	.36	.37	.38	.39	.40
1	.3100	.3200	.3300	.3400	.3500	.3600	.3700	.3800	.3900	.4000
	.41	.42	.43	.44	.45	.46	.47	.48	.49	.50
1	.4100	.4200	.4300	.4400	.4500	.4600	.4700	.4800	.4900	.5000

n = 2

r \ p	.01	.02	.03	.04	.05	.06	.07	.08	.09	.10
1	.0199	.0396	.0591	.0784	.0975	.1164	.1351	.1536	.1719	.1900
2	.0001	.0004	.0009	.0016	.0025	.0036	.0049	.0064	.0081	.0100
	.11	.12	.13	.14	.15	.16	.17	.18	.19	.20
1	.2079	.2256	.2431	.2604	.2775	.2944	.3111	.3276	.3439	.3600
2	.0121	.0144	.0169	.0196	.0225	.0256	.0289	.0324	.0361	.0400
	.21	.22	.23	.24	.25	.26	.27	.28	.29	.30
1	.3759	.3916	.4071	.4224	.4375	.4524	.4671	.4816	.4959	.5100
2	.0441	.0484	.0529	.0576	.0625	.0676	.0729	.0784	.0841	.0900
	.31	.32	.33	.34	.35	.36	.37	.38	.39	.40
1	.5239	.5376	.5511	.5644	.5775	.5904	.6031	.6156	.6279	.6400
2	.0961	.1024	.1089	.1156	.1225	.1296	.1369	.1444	.1521	.1600
	.41	.42	.43	.44	.45	.46	.47	.48	.49	.50
1	.6519	.6636	.6751	.6864	.6975	.7084	.7191	.7296	.7399	.7500
2	.1681	.1764	.1849	.1936	.2025	.2116	.2209	.2304	.2401	.2500

n = 3

r \ p	.01	.02	.03	.04	.05	.06	.07	.08	.09	.10
1	.0297	.0588	.0873	.1153	.1426	.1694	.1956	.2213	.2264	.2710
2	.0003	.0012	.0026	.0047	.0072	.0104	.0140	.0182	.0228	.0280
3	.0000	.0000	.0000	.0001	.0001	.0002	.0003	.0005	.0007	.0010

Table A.1—*Continued*

n = 3 (Continued)

r \ p	.11	.12	.13	.14	.15	.16	.17	.18	.19	.20
1	.2950	.3185	.3415	.3639	.3859	.4073	.4282	.4486	.4686	.4880
2	.0336	.0397	.0463	.0533	.0608	.0686	.0769	.0855	.0946	.1040
3	.0013	.0017	.0022	.0027	.0034	.0041	.0049	.0058	.0069	.0080
	.21	.22	.23	.24	.25	.26	.27	.28	.29	.30
1	.5070	.5254	.5435	.5610	.5781	.5948	.6110	.6268	.6421	.6570
2	.1138	.1239	.1344	.1452	.1562	.1676	.1793	.1913	.2035	.2160
3	.0093	.0106	.0122	.0138	.0156	.0176	.0197	.0220	.0244	.0270
	.31	.32	.33	.34	.35	.36	.37	.38	.39	.40
1	.6715	.6856	.6992	.7125	.7254	.7379	.7500	.7617	.7730	.7840
2	.2287	.2417	.2548	.2682	.2818	.2955	.3094	.3235	.3377	.3520
3	.0298	.0328	.0359	.0393	.0429	.0467	.0507	.0549	.0593	.0640
	.41	.42	.43	.44	.45	.46	.47	.48	.49	.50
1	.7946	.8049	.8148	.8244	.8336	.8425	.8511	.8594	.8673	.8750
2	.3665	.3810	.3957	.4104	.4252	.4401	.4551	.4700	.4850	.5000
3	.0689	.0741	.0797	.0852	.0911	.0973	.1038	.1106	.1176	.1250

n = 4

r \ p	.01	.02	.03	.04	.05	.06	.07	.08	.09	.10
1	.0394	.0776	.1147	.1507	.1855	.2193	.2519	.2836	.3143	.3439
2	.0006	.0023	.0052	.0091	.0140	.0199	.0267	.0344	.0430	.0523
3	.0000	.0000	.0001	.0002	.0005	.0008	.0013	.0019	.0027	.0037
4	.0000	.0000	.0000	.0000	.0000	.0000	.0000	.0000	.0001	.0001
	.11	.12	.13	.14	.15	.16	.17	.18	.19	.20
1	.3726	.4003	.4271	.4530	.4780	.5021	.5254	.5479	.5695	.5904
2	.0624	.0732	.0847	.0968	.1095	.1228	.1366	.1509	.1656	.1808
3	.0049	.0063	.0079	.0098	.0120	.0144	.0171	.0202	.0235	.0272
4	.0001	.0002	.0003	.0004	.0005	.0007	.0008	.0010	.0013	.0016
	.21	.22	.23	.24	.25	.26	.27	.28	.29	.30
1	.6105	.6298	.6485	.6664	.6836	.7001	.7160	.7313	.7459	.7599
2	.1963	.2122	.2285	.2450	.2617	.2787	.2959	.3132	.3307	.3483
3	.0312	.0356	.0403	.0453	.0508	.0566	.0628	.0694	.0763	.0837
4	.0019	.0023	.0028	.0033	.0039	.0046	.0053	.0061	.0071	.0081
	.31	.32	.33	.34	.35	.36	.37	.38	.39	.40
1	.7733	.7862	.7985	.8103	.8215	.8322	.8425	.8522	.8615	.8704
2	.3660	.3837	.4015	.4193	.4370	.4547	.4724	.4900	.5075	.5248
3	.0915	.0996	.1082	.1171	.1265	.1362	.1464	.1569	.1679	.1792
4	.0092	.0105	.0119	.0134	.0150	.0168	.0187	.0209	.0231	.0256
	.41	.42	.43	.44	.45	.46	.47	.48	.49	.50
1	.8788	.8868	.8944	.9017	.9085	.9150	.9211	.9269	.9323	.9375
2	.5420	.5590	.5759	.5926	.6090	.6252	.6412	.6569	.6724	.6875
3	.1909	.2030q	.2155	.2283	.2415	.2550	.2689	.2831	.2977	.3125
4	.0283	.0311	.0342	.0375	.0410	.0448	.0488	.0531	.0576	.0625

n = 5

r \ p	.01	.02	.03	.04	.05	.06	.07	.08	.09	.10
1	.0490	.0961	.1413	.1846	.2262	.2661	.3043	.3409	.3760	.4095
2	.0010	.0038	.0085	.0148	.0226	.0319	.0425	.0544	.0674	.0815
3	.0000	.0001	.0003	.0006	.0012	.0020	.0031	.0045	.0063	.0086
4	.0000	.0000	.0000	.0000	.0000	.0001	.0001	.0002	.0003	.0005
	.11	.12	.13	.14	.15	.16	.17	.18	.19	.20
1	.4416	.4723	.5016	.5296	.5563	.5818	.6061	.6293	.6513	.6723
2	.0965	.1125	.1292	.1467	.1648	.1835	.2027	.2224	.2424	.2627
3	.0112	.0143	.0179	.0220	.0266	.0318	.0375	.0437	.0505	.0579
4	.0007	.0009	.0013	.0017	.0022	.0029	.0036	.0045	.0055	.0067
5	.0000	.0000	.0000	.0001	.0001	.0001	.0001	.0002	.0002	.0003

Table A.1—*Continued*

n = 5 (Continued)

r \ p	.21	.22	.23	.24	.25	.26	.27	.28	.29	.30
1	.6923	.7113	.7293	.7464	.7627	.7781	.7927	.8065	.8196	.8319
2	.2833	.3041	.3251	.3461	.3672	.3883	.4093	.4303	.4511	.4718
3	.0659	.0744	.0836	.0933	.1035	.1143	.1257	.1376	.1501	.1631
4	.0081	.0097	.0114	.0134	.0156	.0181	.0208	.0238	.0272	.0308
5	.0004	.0005	.0006	.0008	.0010	.0012	.0014	.0017	.0021	.0024

r \ p	.31	..32	.33	.34	.35	.36	.37	.38	.39	.40
1	.8436	.8546	.8650	.8748	.8840	.8926	.9008	.9084	.9155	.9222
2	.4923	.5125	.5325	.5522	.5716	.5906	.6093	.6276	.6455	.6630
3	.1766	.1905	.2050	.2199	.2352	.2509	.2670	.2835	.3003	.3174
4	.0347	.0390	.0436	.0486	.0540	.0598	.0660	.0726	.0796	.0870
5	.0029	.0034	.0039	.0045	.0053	.0060	.0069	.0079	.0090	.0102

r \ p	.41	.42	.43	.44	.45	.46	.47	.48	.49	.50
1	.9285	.9344	.9398	.9449	.9497	.9541	.9582	.9620	.9655	.9688
2	.6801	.6967	.7129	.7286	.7438	.7585	.7728	.7865	.7998	.8125
3	.3349	.3525	.3705	.3886	.4069	.4253	.4439	.4625	.4813	.5000
4	.0949	.1033	.1121	.1214	.1312	.1415	.1522	.1635	.1752	.1875
5	.0116	.0131	.0147	.0165	.0185	.0206	.0229	.0255	.0282	.0312

n = 6

r \ p	.01	.02	.03	.04	.05	.06	..07	.08	.09	.10
1	.0585	.1142	.1670	.2172	.2649	.3101	.3530	.3936	.4321	.4686
2	.0015	.0057	.0125	.0216	.0328	.0459	.0608	.0773	.0952	.1143
3	.0000	.0002	.0005	.0012	.0022	.0038	.0058	.0085	.0118	.0158
4	.0000	.0000	.0000	.0000	.0001	.0002	.0003	.0005	.0008	.0013
5	.0000	.0000	.0000	.0000	.0000	.0000	.0000	.0000	.0000	.0001

r \ p	.11	.12	.13	.14	.15	.16	.17	.18	.19	.20
1	.5030	.5356	.5664	.5954	.6229	.6487	.6731	.6960	.7176	.7379
2	.1345	.1556	.1776	.2003	.2235	.2472	.2713	.2956	.3201	.3446
3	.0206	.0261	.0324	.0395	.0476	.0560	.0655	.0759	.0870	.0989
4	.0018	.0025	.0034	.0045	.0059	.0075	.0094	.0116	.0141	.0170
5	.0001	.0001	.0002	.0003	.0004	.0005	.0007	.0010	.0013	.0016
6	.0000	.0000	.0000	.0000	.0000	.0000	.0000	.0000	.0000	.0001

r \ p	.21	.22	.23	.24	.25	.26	.27	.28	.29	.30
1	.7569	.7748	.7916	.8073	.8220	.8358	.8487	.8607	.8719	.8824
2	.3692	.3937	.4180	.4422	.4661	.4896	.5128	.5356	.5580	.5798
3	.1115	.1250	.1391	.1539	.1694	.1856	.2023	.2196	.2374	.2557
4	.0202	.0239	.0280	.0326	.0376	.0431	.0492	.0557	.0628	.0705
5	.0020	.0025	.0031	.0038	.0046	.0056	.0067	.0079	.0093	.0109
6	.0001	.0001	.0001	.0002	.0002	.0003	.0004	.0005	.0006	.0007

r \ p	.31	.32	.33	.34	.35	.36	.37	.38	.39	.40
1	.8921	.9011	.9095	.9173	.9246	.9313	.9375	.9432	.9485	.9533
2	.6012	.6220	.6422	.6619	.6809	.6994	.7172	.7343	.7508	.7667
3	.2744	.2936	.3130	.3328	.3529	.3732	.3937	.4143	.4350	.4557
4	.0787	.0875	.0969	.1069	.1174	.1286	.1404	.1527	.1657	.1792
5	.0127	.0148	.0170	.0195	.0223	.0254	.0288	.0325	.0365	.0410
6	.0009	.0011	.0013	.0015	.0018	.0022	.0026	.0030	.0035	.0041

r \ p	.41	.42	.43	.44	.45	.46	.47	.48	.49	.50
1	.9578	.9619	.9657	.9692	.9723	.9752	.9778	.9802	.9824	.9844
2	.7819	.7965	.8105	.8238	.8364	.8485	.8599	.8707	.8810	.8906
3	.4764	.4971	.5177	.5382	.5585	.5786	.5985	.6180	.6373	.6562
4	.1933	.2080	.2232	.2390	.2553	.2721	.2893	.3070	.3252	.3438
5	.0458	.0510	.0566	.0627	.0692	.0762	.0837	.0917	.1003	.1094
6	.0048	.0055	.0063	.0073	.0083	.0095	.0108	.0122	.0138	.0156

Table A.1—*Continued*

n = 7

r \ p	.01	.02	.03	.04	.05	.06	.07	.08	.09	.10
1	.0679	.1319	.1920	.2486	.3017	.3515	.3983	.4422	.4832	.5217
2	.0020	.0079	.0171	.0294	.0444	.0618	.0813	.1026	.1255	.1497
3	.0000	.0003	.0009	.0020	.0038	.0063	.0097	.0140	.0193	.0257
4	.0000	.0000	.0000	.0001	.0002	.0004	.0007	.0012	.0018	.0027
5	.0000	.0000	.0000	.0000	.0000	.0000	.0000	.0001	.0001	.0002
	.11	.12	.13	.14	.15	.16	.17	.18	.29	.20
1	.5577	.5913	.6227	.6521	.6794	.7049	.7286	.7507	.7712	.7903
2	.1750	.2012	.2281	.2556	.2834	.3115	.3396	.3677	.3956	.4233
3	.0331	.0416	.0513	.0620	.0738	.0866	.1005	.1154	.1313	.1480
4	.0039	.0054	.0072	.0094	.0121	.0153	.0189	.0231	.0279	.0333
5	.0003	.0004	.0006	.0009	.0012	.0017	.0022	.0029	.0037	.0047
6	.0000	.0000	.0000	.0000	.0001	.0001	.0001	.0002	.0003	.0004
	.21	.22	.23	.24	.25	.26	.27	.28	.29	.30
1	.8080	.8243	.8395	.8535	.8665	.8785	.8895	.8997	.9090	.9176
2	.4506	.4775	.5040	.5298	.5551	.5796	.6035	.6266	.6490	.6706
3	.1657	.1841	.2033	.2231	.2436	.2646	.2861	.3081	.3304	.3529
4	.0394	.0461	.0536	.0617	.0706	.0802	.0905	.1016	.1134	.1260
5	.0058	.0072	.0088	.0107	.0129	.0153	.0181	.0213	.0248	.0288
6	.0005	.0006	.0008	.0011	.0013	.0017	.0021	.0026	.0031	.0038
7	.0000	.0000	.0000	.0000	.0001	.0001	.0001	.0001	.0002	.0002
	.31	.32	.33	.34	.35	.36	.37	.38	.39	.40
1	.9255	.9328	.9394	.9454	.9510	.9560	.9606	.9648	.9686	.9720
2	.6914	.7113	.7304	.7487	.7662	.7828	.7987	.8137	.8279	.8414
3	.3757	.3987	.4217	.4447	.4677	.4906	.5134	.5359	.5581	.5801
4	.1394	.1534	.1682	.1837	.1998	.2167	.2341	.2521	.2707	.2898
5	.0332	.0380	.0434	.0492	.0556	.0625	.0701	.0782	.0869	.0963
6	.0046	.0055	.0065	.0077	.0090	.0105	.0123	.0142	.0164	.0188
7	.0003	.0003	.0004	.0005	.0006	.0008	.0009	.0011	.0014	.0016
	.41	.42	.43	.44	.45	.46	.47	.48	.49	.50
1	.9751	.9779	.9805	.9827	.9848	.9866	.9883	.9897	.9910	.9922
2	.8541	.8660	.8772	.8877	.8976	.9068	.9153	.9233	.9307	.9375
3	.6017	.6229	.6436	.6638	.6836	.7027	.7213	.7393	.7567	.7734
4	.3094	.3294	.3498	.3706	.3917	.4131	.4346	.4563	.4781	.5000
5	.1063	.1169	.1282	.1402	.1529	.1663	.1803	.1951	.2105	.2266
6	.0216	.0246	.0279	.0316	.0357	.0402	.0451	.0504	.0562	.0625
7	.0019	.0023	.0027	.0032	.0037	.0044	.0051	.0059	.0068	.0078

n = 8

r \ p	.01	.02	.03	.04	.05	.06	.07	.08	.09	.10
1	.0773	.1492	.2163	.2786	.3366	.3904	.4404	.4868	.5297	.5695
2	.0027	.0103	.0223	.0381	.0572	.0792	.1035	.1298	.1577	.1869
3	.0001	.0004	.0013	.0031	.0058	.0096	.0147	.0211	.0289	.0381
4	.0000	.0000	.0001	.0002	.0004	.0007	.0013	.0022	.0034	.0050
5	.0000	.0000	.0000	.0000	.0000	.0000	.0001	.0001	.0003	.0004
	.11	.12	.13	.14	.15	.16	.17	.18	.19	.20
1	.6063	.6404	.6718	.7008	.7275	.7521	.7748	.7956	.8147	.8322
2	.2171	.2480	.2794	.3111	.3428	.3744	.4057	.4366	.4670	.4967
3	.0487	.0608	.0743	.0891	.1052	.1226	.1412	.1608	.1815	.2031
4	.0071	.0097	.0129	.0168	.0214	.0267	.0328	.0397	.0476	.0563
5	.0007	.0010	.0015	.0021	.0029	.0038	.0050	.0065	.0083	.0104
6	.0000	.0001	.0001	.0002	.0002	.0004	.0005	.0007	.0009	.0012
7	.0000	.0000	.0000	.0000	.0000	.0000	.0000	.0000	.0001	.0001
	.21	.22	.23	.24	.25	.26	.27	.28	.29	.30
1	.8483	.8630	.8764	.8887	.8999	.9101	.9194	.9278	.9354	.9424
2	.5257	.5538	.5811	.6075	.6329	.6573	.6807	.7031	.7244	.7447
3	.2255	.2486	.2724	.2967	.3215	.3465	.3718	.3973	.4228	.4482
4	.0659	.0765	.0880	.1004	.1138	.1281	.1433	.1594	.1763	.1941
5	.0129	.0158	.0191	.0230	.0273	.0322	.0377	.0438	.0505	.0580
6	.0016	.0021	.0027	.0034	.0042	.0052	.0064	.0078	.0094	.0113
7	.0001	.0002	.0002	.0003	.0004	.0005	.0006	.0008	.0010	.0013
8	.0000	.0000	.0000	.0000	.0000	.0000	.0000	.0000	.0001	.0001

Table A.1—*Continued*

n = 8 (Continued)

r \ p	.31	.32	.33	.34	.35	.36	.37	.38	.39	.40
1	.9486	.9543	.9594	.9640	.9681	.9719	.9752	.9782	.9808	.9832
2	.7640	.7822	.7994	.8156	.8309	.8452	.8586	.8711	.8828	.8936
3	.4736	.4987	.5236	.5481	.5722	.5958	.6189	.6415	.6634	.6846
4	.2126	.2319	.2519	.2724	.2936	.3153	.3374	.3599	.3828	.4059
5	.0661	.0750	.0846	.0949	.1061	.1180	.1307	.1443	.1586	.1737
6	.0134	.0159	.0187	.9218	.0253	.0293	.0336	.0385	.0439	.0498
7	.0016	.0020	.0024	.0030	.0036	.0043	.0051	.0061	.0072	.0085
8	.0001	.0001	.0001	.0002	.0002	.0003	.0004	.0004	.0005	.0007

r \ p	.41	.42	.43	.44	.45	.46	.47	.48	.49	.50
1	.9853	.9872	.9889	.9903	.9916	.9928	.9938	.9947	.9954	.9961
2	.9037	.9130	.9216	.9295	.9368	.9435	.9496	.9552	.9602	.9648
3	.7052	.7250	.7440	.7624	.7799	.7966	.8125	.8276	.8419	.8555
4	.4292	.4527	.4762	.4996	.5230	.5463	.5694	.5922	.6146	.6367
5	.1895	.2062	.2235	.2416	.2604	.2798	.2999	.3205	.3416	.3633
6	.0563	.0634	.0711	.0794	.0885	.0982	.1086	.1198	.1318	.1445
7	.0100	.0117	.0136	.0157	.0181	.0208	.0239	.0272	.0310	.0352
8	.0008	.0010	.0012	.0014	.0017	.0020	.0024	.0028	.0033	.0039

n = 9

r \ p	.01	.02	.03	.04	.05	.06	.07	.08	.09	.10
1	.0865	.1663	.2398	.3075	.3698	.4270	.4796	.5278	.5721	.6126
2	.0034	.0131	.0282	.0478	.0712	.0978	.1271	.1583	.1912	.2252
3	.0001	.0006	.0020	.0045	.0084	.0138	.0209	.0298	.0405	.0530
4	.0000	.0000	.0001	.0003	.0006	.0013	.0023	.0037	.0057	.0083
5	.0000	.0000	.0000	.0000	.0000	.0001	.0002	.0003	.0005	.0009
6	.0000	.0000	.0000	.0000	.0000	.0000	.0000	.0000	.0000	.0001

r \ p	.11	.12	.13	.14	.15	.16	.17	.18	.19	.20
1	.6496	.6835	.7145	.7427	.7684	.7918	.8131	.8324	.8499	.8658
2	.2599	.2951	.3304	.3657	.4005	.4348	.4685	.5012	.5330	.5638
3	.0672	.0833	.1009	.1202	.1409	.1629	.1861	.2105	.2357	.2618
4	.0117	.0158	.0209	.0269	.0339	.0420	.0512	.0615	.0730	.0856
5	.0014	.0021	.0030	.0041	.0056	.0075	.0098	.0125	.0158	.0196
6	.0001	.0002	.0003	.0004	.0006	.0009	.0013	.0017	.0023	.0031
7	.0000	.0000	.0000	.0000	.0000	.0001	.0001	.0002	.0002	.0003

r \ p	.21	.22	.23	.24	.25	.26	.27	.28	.29	.30
1	.8801	.8931	.9048	.9154	.9249	.9335	.9411	.9480	.9542	.9596
2	.5934	.6218	.6491	.6750	.6997	.7230	.7452	.7660	.7856	.8040
3	.2885	.3158	.3434	.3713	.3993	.4273	.4552	.4829	.5102	.5372
4	.0994	.0114	.1304	.1475	.1657	.1849	.2050	.2260	.2478	.2703
5	.0240	.0291	.0350	.0416	.0489	.0571	.0662	.0762	.0870	.0988
6	.0040	.0051	.0065	.0081	.0100	.0122	.0149	.0179	.0213	.0253
7	.0004	.0006	.0008	.0010	.0013	.0017	.0022	.0028	.0035	.0043
8	.0000	.0000	.0001	.0001	.0001	.0001	.0002	.0003	.0003	.0004

r \ p	.31	.32	.33	.34	.35	.36	.37	.38	.39	.40
1	.9645	.9689	.9728	.9762	.9793	.9820	.9844	.9865	.9883	.9899
2	.8212	.8372	.8522	.8661	.8789	.8908	.9017	.9118	.9210	.9295
3	.5636	.5894	.6146	.6390	.6627	.6856	.7076	.7287	.7489	.7682
4	.2935	.3173	.3415	.3662	.3911	.4163	.4416	.4669	.4922	.5174
5	.1115	.1252	.1398	.1553	.1717	.1890	.2072	.2262	.2460	.2666
6	.0298	.0348	.0404	.0467	.0536	.0612	.0696	.0787	.0886	.0994
7	.0053	.0064	.0078	.0094	.0112	.0133	.0157	.0184	.0215	.0250
8	.0006	.0007	.0009	.0011	.0014	.0017	.0021	.0026	.0031	.0038
9	.0000	.0000	.0000	.0001	.0001	.0001	.0001	.0002	.0002	.0003

r \ p	.41	.42	.43	.44	.45	.46	.47	.48	.49	.50
1	.9913	.9926	.9936	.9946	.9954	.9961	.9967	.9972	.9977	.9980
2	.9372	.9442	.9505	.0563	.9615	.9662	.9704	.9741	.9775	.9805
3	.7866	.8039	.8204	.8359	.8505	.8642	.8769	.8889	.8999	.9102
4	.5424	.5670	.5913	.6152	.6386	.6614	.6836	.7052	.7260	.7461
5	.2878	.3097	.3322	.3551	.3786	.4024	.4265	.4509	.4754	.5000
6	.1109	.1233	.1366	.1508	.1658	.1817	.1985	.2161	.2346	.2539
7	.0290	.0334	.0383	.0437	.0498	.0564	.0637	.0717	.0804	.0898
8	.0046	.0055	.0065	.0077	.0091	.0107	.0125	.0145	.0169	.0195
9	.0003	.0004	.0005	.0006	.0008	.0009	.0011	.0014	.0016	.0020

Table A.1—*Continued*

n = 10

r \ p	.01	.02	.03	.04	.05	.06	.07	.08	.09	.10
1	.0956	.1829	.2626	.3352	.4013	.4614	.5160	.5656	.6106	.6513
2	.0043	.0162	.0345	.0582	.0861	.1176	.1517	.1879	.2254	.2639
3	.0001	.0009	.0028	.0062	.0115	.0188	.0283	.0401	.0540	.0702
4	.0000	.0000	.0001	.0004	.0010	.0020	.0036	.0058	.0088	.0128
5	.0000	.0000	.0000	.0000	.0001	.0002	.0003	.0006	.0010	.0016
6	.0000	.0000	.0000	.0000	.0000	.0000	.0000	.0000	.0001	.0001
	.11	.12	.13	.14	.15	.16	.17	.18	.19	.20
1	.6882	.7215	.7516	.7787	.8031	.8251	.8448	.8626	.8784	.8926
2	.3028	.3417	.3804	.4184	.4557	.4920	.5270	.5608	.5932	.6242
3	.0884	.1087	.1308	.1545	.1798	.2064	.2341	.2628	.2922	.3222
4	.0178	.0239	.0313	.0400	.0500	.0614	.0741	.0883	.1039	.1209
5	.0025	.0037	.0053	.0073	.0099	.0130	.0168	.0213	.0266	.0328
6	.0003	.0004	.0006	.0010	.0014	.0020	.0027	.0037	.0049	.0064
7	.0000	.0000	.0001	.0001	.0001	.0002	.0003	.0004	.0006	.0009
8	.0000	.0000	.0000	.0000	.0000	.0000	.0000	.0000	.0001	.0001
	.21	.22	.23	.24	.25	.26	.27	.28	.29	.30
1	.9053	.9166	.9267	.9357	.9437	.9508	.9570	.9626	.9674	.9718
2	.6536	.6815	.7079	.7327	.7560	.7778	.7981	.8170	.8345	.8507
3	.3526	.3831	.4137	.4442	.4744	.5042	.5335	.5622	.5901	.6172
4	.1391	.1587	.1794	.2012	.2241	.2479	.2726	.2979	.3239	.3504
5	.0399	.0479	.0569	.0670	.0781	.0904	.1037	.1181	.1337	.1503
6	.0082	.0104	.0130	.0161	.0197	.0239	.0287	.0342	.0404	.0473
7	.0012	.0016	.0021	.0027	.0035	.0045	.0056	.0070	.0087	.0106
8	.0001	.0002	.0002	.0003	.0004	.0006	.0007	.0010	.0012	.0016
9	.0000	.0000	.0000	.0000	.0000	.0000	0001	.0001	.0001	.0001
	.31	.32	.33	.34	.35	.36	.37	.38	.39	.40
1	.9755	.9789	.9818	.9843	.9865	.9885	.9902	.9916	.9929	.9940
2	.8656	.8794	.8920	.9035	.9140	.9236	.9323	.9402	.9473	.9536
3	.6434	.6687	.6930	.7162	.7384	.7595	.7794	.7983	.8160	.8327
4	.3772	.4044	.4316	.4589	.4862	.5132	.5400	.5664	.5923	.6177
5	.1679	.1867	.2064	.2270	.2485	.2708	.2939	.3177	.3420	.3669
6	.0551	.0637	.0732	.0836	.0949	.1072	.1205	.1348	.1500	.1662
7	.0129	.0155	.0185	.0220	.0260	.0305	.0356	.0413	.0477	.0548
8	.0020	.0025	.0032	.0039	.0048	.0059	.0071	.0086	.0103	.0123
9	.0002	.0003	.0003	.0004	.0005	.0007	.0009	.0011	.0014	.0017
10	.0000	.0000	.0000	.0000	.0000	.0000	.0000	.0001	.0001	.0001
	.41	.42	.43	.44	.45	.46	.47	.48	.49	.50
1	.9949	.9957	.9964	.9970	.9975	.9979	.9983	.9986	.9988	.9990
2	.9594	.9645	.9691	.9731	.9767	.9799	.9827	.9852	.9874	.9892
3	.8483	.8628	.8764	.8889	.9004	.9111	.9209	.9298	.9379	.9453
4	.6425	.6665	.6898	.7123	.7340	.7547	.7745	.7933	.8112	.8281
5	.3922	.4178	.4436	.4696	.4956	.5216	.5474	.5730	.5982	.6230
6	.1834	.2016	.2207	.2407	.2616	.2832	.3057	.3288	.3526	.3770
7	.0626	.0712	.0806	.0908	.1020	.1141	.1271	.1410	.1560	.1719
8	.0146	.0172	.0202	.0236	.0274	.0317	.0366	.0420	.0480	.0547
9	.0021	.0025	.0031	.0037	.0045	.0054	.0065	.0077	.0091	.0107
10	.0001	.0002	.0002	.0003	.0003	.0004	.0005	.0006	.0008	.0010

n = 11

r \ p	.01	.02	.03	.04	.05	.06	.07	.08	.09	.10
1	.1047	.1993	.2847	.3618	.4312	.4937	.5499	.6004	.6456	.6862
2	.0052	.0195	.0413	.0692	.1019	.1382	.1772	.2181	.2601	.3026
3	.0002	.0012	.0037	.0083	.0152	.0248	.0370	.0519	.0695	.0896
4	.0000	.0000	.0002	.0007	.0016	.0030	.0053	.0085	.0129	.0185
5	.0000	.0000	.0000	.0000	.0001	.0003	.0005	.0010	.0017	.0028
6	.0000	.0000	.0000	.0000	.0000	.0000	.0000	.0001	.0002	.0003
	.11	.12	.13	.14	.15	.16	.17	.18	.19	.20
1	.7225	.7549	.7839	.8097	.8327	.8531	.8712	.8873	.9015	.9141
2	.3452	.3873	.4286	.4689	.5078	.5453	.5811	.6151	.6474	.6779
3	.1120	.1366	.1632	.1915	.2212	.2521	.2839	.3164	.3494	.3826
4	.0256	.0341	.0442	.0560	.0694	.0846	.1013	.1197	.1397	.1611
5	.0042	.0061	.0087	.0119	.0159	.0207	.0266	.0334	.0413	.0504
6	.0005	.0008	.0012	.0018	.0027	.0037	.0051	.0068	.0090	.0117
7	.0000	.0001	.0001	.0002	.0003	.0005	.0007	.0010	.0014	.0020
8	.0000	.0000	.0000	.0000	.0000	.0000	.0001	.0001	.0002	.0002

Table A.1—*Continued*

n = 11 (Continued)

r \ p	.21	.22	.23	.24	.25	.26	.27	.28	.29	.30
1	.9252	.9350	.9436	.9511	.9578	.9636	.9686	.9730	.9769	.9802
2	.7065	.7333	.7582	.7814	.8029	.8227	.8410	.8577	.8730	.8870
3	.4158	.4488	.4814	.5134	.5448	.5753	.6049	.6335	.6610	.6873
4	.1840	.2081	.2333	.2596	.2867	.3146	.3430	.3719	.4011	.4304
5	.0607	.0723	.0851	.0992	.1146	.1313	.1493	.1685	.1888	.2103
6	.0148	.0186	.0231	.0283	.0343	.0412	.0490	.0577	.0674	.0782
7	.0027	.0035	.0046	.0059	.0076	.0095	.0119	.0146	.0179	.0216
8	.0003	.0005	.0007	.0009	.0012	.0016	.0021	.0027	.0034	.0043
9	.0000	.0000	.0001	.0001	.0001	.0002	.0002	.0003	.0004	.0006
	.31	.32	.33	.34	.35	.36	.37	.38	.39	.40
1	.9831	.9856	.9878	.9896	.9912	.9926	.9938	.9948	.9956	.9964
2	.8997	.9112	.9216	.9310	.9394	.9470	.9537	.9597	.9650	.9698
3	.7123	.7361	.7587	.7799	.7999	.8186	.8360	.8522	.8672	.8811
4	.4598	.4890	.5179	.5464	.5744	.6019	.6286	.6545	.6796	.7037
5	.2328	.2563	.2807	.3059	.3317	.3581	.3850	.4122	.4397	.4672
6	.0901	.1031	.1171	.1324	.1487	.1661	.1847	.2043	.2249	.2465
7	.0260	.0309	.0366	.0430	.0501	.0581	.0670	.0768	.0876	.0994
8	.0054	.0067	.0082	.0101	.0122	.0148	.0177	.0210	.0249	.0293
9	.0008	.0010	.0013	.0016	.0020	.0026	.0032	.0039	.0048	.0059
10	.0001	.0001	.0001	.0002	.0002	.0003	.0004	.0005	.0006	.0007
	.41	.42	.43	.44	.45	.46	.47	.48	.49	.50
1	.9970	.9975	.9979	.9983	.9986	.9989	.9991	.9992	.9994	.9995
2	.9739	.9776	.9808	.9836	.9861	.9882	.9900	.9916	.9930	.9941
3	.8938	.9055	.9162	.9260	.9348	.9428	.9499	.9564	.9622	.9673
4	.7269	.7490	.7700	.7900	.8089	.8266	.8433	.8588	.8733	.8867
5	.4948	.5223	.5495	.5764	.6029	.6288	.6541	.6787	.7026	.7256
6	.2690	.2924	.3166	.3414	.3669	.3929	.4193	.4460	.4729	.5000
7	.1121	.1260	.1408	.1568	.1738	.1919	.2110	.2312	.2523	.2744
8	.0343	.0399	.0461	.0532	.0610	.0696	.0791	.0895	.1009	.1133
9	.0072	.0087	.0104	.0125	.0148	.0175	.0206	.0241	.0282	.0327
10	.0009	.0012	.0014	.0018	.0022	.0027	.0033	.0040	.0049	.0059
11	.0001	.0001	.0001	.0001	.0002	.0002	.0002	.0003	.0004	.0005

n = 12

r \ p	.01	.02	.03	.04	.05	.06	.07	.08	.09	.10
1	.1136	.2153	.3062	.3873	.4596	.5241	.5814	.6323	.6775	.7176
2	.0062	.0231	.0486	.0809	.1184	.1595	.2033	.2487	.2948	.3410
3	.0002	.0015	.0048	.0107	.0196	.0316	.0468	.0652	.0866	.1109
4	.0000	.0001	.0003	.0010	.0022	.0043	.0075	.0120	.0180	.0256
5	.0000	.0000	.0000	.0001	.0002	.0004	.0009	.0016	.0027	.0043
6	.0000	.0000	.0000	.0000	.0000	.0000	.0001	.0002	.0003	.0005
7	.0000	.0000	.0000	.0000	.0000	.0000	.0000	.0000	.0000	.0001
	.11	.12	.13	.14	.15	.16	.17	.18	.19	.20
1	.7530	.7843	.8120	.8363	.8578	.8766	.8931	.9076	.9202	.9313
2	.3867	.4314	.4748	.5166	.5565	.5945	.6304	.6641	.6957	.7251
3	.1377	.1667	.1977	.2303	.2642	.2990	.3344	.3702	.4060	.4417
4	.0351	.0464	.0597	.0750	.0922	.1114	.1324	.1552	.1795	.2054
5	.0065	.0095	.0133	.0181	.0239	.0310	.0393	.0489	.0600	.0726
6	.0009	.0014	.0022	.0033	.0046	.0065	.0088	.0116	.0151	.0194
7	.0001	.0002	.0003	.0004	.0007	.0010	.0015	.0021	.0029	.0039
8	.0000	.0000	.0000	.0000	.0001	.0001	.0002	.0003	.0004	.0006
9	.0000	.0000	.0000	.0000	.0000	.0000	.0000	.0000	.0000	.0001
	.21	.22	.23	.24	.25	.26	.27	.28	.29	.30
1	.9409	.9493	.9566	.9629	.9683	.9730	.9771	.9806	.9836	.9862
2	.7524	.7776	.8009	.8222	.8416	.8594	.8755	.8900	.9032	.9150
3	.4768	.5114	.5450	.5778	.6093	.6397	.6687	.6963	.7225	.7472
4	.2326	.2610	.2904	.3205	.3512	.3824	.4137	.4452	.4765	.5075
5	.0866	.1021	.1192	.1377	.1576	.1790	.2016	.2254	.2504	.2763
6	.0245	.0304	.0374	.0453	.0544	.0646	.0760	.0887	.1026	.1178
7	.0052	.0068	.0089	.0113	.0143	.0178	.0219	.0267	.0322	.0386
8	.0008	.0011	.0016	.0021	.0028	.0036	.0047	.0060	.0076	.0095
9	.0001	.0001	.0002	.0003	.0004	.0005	.0007	.0010	.0013	.0017
10	.0000	.0000	.0000	.0000	.0000	.0001	.0001	.0001	.0002	.0002

Table A.1—*Continued*

n = 12 (Continued)

r \ p	.31	.32	.33	.34	.35	.36	.37	.38	.39	.40
1	.9884	.9902	.9918	.9932	.9943	.9953	.9961	.9968	.9973	.9978
2	.9256	.9350	.9435	.9509	.9576	.9634	.9685	.9730	.9770	.9804
3	.7704	.7922	.8124	.8313	.8487	.8648	.9795	.8931	.9054	.9166
4	.5381	.5681	.5973	.6258	.6533	.6799	.7053	.7296	.7528	.7747
5	.3032	.3308	.3590	.3876	.4167	.4459	.4751	.5043	.5332	.5618
6	.1343	.1521	.1711	.1913	.2127	.2352	.2588	.2833	.3087	.3348
7	.0458	.0540	.0632	.0734	.0846	.0970	.1106	.1253	.1411	.1582
8	.0118	.0144	.0176	.0213	.0255	.0304	.0359	.0422	.0493	.0573
9	.0022	.0028	.0036	.0045	.0056	.0070	.0086	.0104	.0127	.0153
10	.0003	.0004	.0005	.0007	.0008	.0011	.0014	.0018	.0022	.0028
11	.0000	.0000	.0000	.0001	.0001	.0001	.0001	.0002	.0002	.0003

r \ p	.41	.42	.43	.44	.45	.46	.47	.48	.49	.50
1	.9982	.9986	.9988	.9990	.9992	.9994	.9995	.9996	.9997	.9998
2	.9834	.9860	.9882	.9901	.9917	.9931	.9943	.9953	.9961	.9968
3	.9267	.9358	.9440	.9513	.9579	.9637	.9688	.9733	.9773	.9807
4	.7953	.8147	.8329	.8498	,7555	.8801	.8934	.9057	.9168	.9270
5	.5899	.6175	.6443	.6704	.6956	.7198	.7430	.7652	.7862	.8062
6	.3616	.3889	.4167	.4448	.4731	.5014	.5297	.5577	.5855	.6128
7	.1765	.1959	.2164	.2380	.2607	.2843	.3089	.3343	.3604	.3872
8	.0662	.0760	.0869	.0988	.1117	.1258	.1411	.1575	.1751	.1938
9	.0183	.0218	.0258	.0304	.0356	.0415	.0481	.0555	.0638	.0730
10	.0035	.0043	.0053	.0065	.0079	.0095	.0114	.0137	.0163	.0193
11	.0004	.0005	.0007	.0009	.0011	.0014	.0017	.0021	.0026	.0032
12	.0000	.0000	.0000	.0001	.0001	.0001	.0001	.0001	.0002	.0002

n = 13

r \ p	.01	.02	.03	.04	.05	.06	.07	.08	.09	.10
1	.1225	.2310	.3270	.4118	.4867	.5526	.6107	.6617	.7065	.7458
2	.0072	.0270	.0564	.0932	.1354	.1814	.2298	.2794	.3293	.3787
3	.0003	.0020	.0062	.0135	.0245	.0392	.0578	.0799	.1054	.1339
4	.0000	.0001	.0005	.0014	.0031	.0060	.0103	.0163	.0242	.0342
5	.0000	.0000	.0000	.0001	.0003	.0007	.0013	.0024	.0041	.0065
6	.0000	.0000	.0000	.0000	.0000	.0001	.0001	.0003	.0005	.0009
7	.0000	.0000	.0000	.0000	.0000	.0000	.0000	.0000	.0001	.0001

r \ p	.11	.12	.13	.14	.15	.16	.17	.18	.19	.20
1	.7802	.8102	.8364	.8592	.8791	.8963	.9113	.9242	.9354	.9450
2	.4270	.4738	.5186	.5614	.6017	.6396	.6751	.7080	.7384	.7664
3	.1651	.1985	.2337	.2704	.3080	.3463	.3848	.4231	.4611	.4983
4	.0464	.0609	.0776	.0967	.1180	.1414	.1667	.1939	.2226	.2527
5	.0097	.0139	.0193	.0260	.0342	.0438	.0551	.0681	.0827	.0991
6	.0015	.0024	.0036	.0053	.0075	.0104	.0139	.0183	.0237	.0300
7	.0002	.0003	.0005	.0008	.0013	.0019	.0027	.0038	.0052	.0070
8	.0000	.0000	.0001	.0001	.0002	.0003	.0004	.0006	.0009	.0012
9	.0000	.0000	.0000	.0000	.0000	.0000	.0000	.0001	.0001	.0002

r \ p	.21	.22	.23	.24	.25	.26	.27	.28	.29	.30
1	.9533	.9604	.9666	.9718	.9762	.9800	.9833	.9860	.9883	.9903
2	.7920	.8154	.8367	.8559	.8733	.8889	.9029	.9154	.9265	.9363
3	.5347	.5699	.6039	.6364	.6674	.6968	.7245	.7505	.7749	.7975
4	.2839	.3161	.3489	.3822	.4157	.4493	.4826	.5155	.5478	.5794
5	.1173	.1371	.1585	.1816	.2060	.2319	.2589	.2870	.3160	.3457
6	.0375	.0462	.0562	.0675	.0802	.0944	.1099	.1270	.1455	.1654
7	.0093	.0120	.0154	.0195	.0243	.0299	.0365	.0440	.0527	.0624
8	.0017	.0024	.0032	.0043	.0056	.0073	.0093	.0118	.0147	.0182
9	.0002	.0004	.0005	.0007	.0010	.0013	.0018	.0024	.0031	.0040
10	.0000	.0000	.0001	.0001	.0001	.0002	.0003	.0004	.0005	.0007
11	.0000	.0000	.0000	.0000	.0000	.0000	.0000	.0000	.0001	.0001

r \ p	.31	.32	.33	.34	.35	.36	.37	.38	.39	.40
1	.9920	.9934	.9945	.9955	.9963	.9970	.9975	.9980	.9984	.9987
2	.9450	.9527	.9594	.9653	.9704	.0749	.9787	.9821	.9849	.9874
3	.8185	.8379	.8557	.8720	.8868	.9003	.9125	.9235	.9333	.9421
4	.6101	.6398	.6683	.6957	.7217	.7464	.7698	.7917	.8123	.8314
5	.3760	.4067	.4376	.4686	.4995	.5301	.5603	.5899	.6188	.6470

Table A.1—*Continued*

n = 13 (Continued)

r \ p	.31	.32	.33	.34	.35	.36	.37	.38	.39	.40
6	.1867	.2093	.2331	.2581	.2841	.3111	.3388	.3673	.3962	.4256
7	.0733	.0854	.0988	.1135	.1295	.1468	.1654	.1853	.2065	.2288
8	.0223	.0271	.0326	.0390	.0462	.0544	.0635	.0738	.0851	.0977
9	.0052	.0065	.0082	.0102	.0126	.0154	.0187	.0225	.0270	.0321
10	.0009	.0012	.0015	.0020	.0025	.0032	.0040	.0051	.0063	.0078
11	.0001	.0001	.0002	.0003	.0003	.0005	.0006	.0008	.0010	.0013
12	.0000	.0000	.0000	.0000	.0000	.0000	.0001	.0001	.0001	.0001

r \ p	.41	.42	.43	.44	.45	.46	.47	.48	.49	.50
1	.9990	.9992	.9993	.9995	.9996	.9997	.9997	.9998	.9998	.9999
2	.9895	.9912	.9928	.9940	.9951	.9960	.9967	.9974	.9979	.9983
3	.9499	.9569	.9630	.9684	.9731	.9772	.9808	.9838	.9865	.9888
4	.8492	.8656	.8807	.8945	.9071	.9185	.9288	.9381	.9464	.9539
5	.6742	.7003	.7254	.7493	.7721	.7935	.8137	.8326	.8502	.8666
6	.4552	.4849	.5146	.5441	.5732	.6019	.6299	.6573	.6838	.7095
7	.2524	.2770	.3025	.3290	.3563	.3842	.4127	.4415	.4707	.5000
8	.1114	.1264	.1426	.1600	.1788	.1988	.2200	.2424	.2659	.2905
9	.0379	.0446	.0520	.0605	.0698	.0803	.0918	.1045	.1183	.1334
10	.0096	.0117	.0141	.0170	.0203	.0242	.0287	.0338	.0396	.0461
11	.0017	.0021	.0027	.0033	.0041	.0051	.0063	.0077	.0093	.0112
12	.0002	.0002	.0003	.0004	.0005	.0007	.0009	.0011	.0014	.0017
13	.0000	.0000	.0000	.0000	.0000	.0000	.0001	.0001	.0001	.0001

n = 14

r \ p	.01	.02	.03	.04	.05	.06	.07	.08	.09	.10
1	.1313	.2464	.3472	.4353	.5123	.5795	.6380	.6888	.7330	.7712
2	.0084	.0310	.0645	.1059	.1530	.2037	.2564	.3100	.3632	.4154
3	.0003	.0025	.0077	.0167	.0301	.0478	.0698	.0958	.1255	.1584
4	.0000	.0001	.0006	.0019	.0042	.0080	.0136	.0214	.0315	.0441
5	.0000	.0000	.0000	.0002	.0004	.0010	.0020	.0035	.0059	.0092
6	.0000	.0000	.0000	.0000	.0000	.0001	.0002	.0004	.0008	.0015
7	.0000	.0000	.0000	.0000	.0000	.0000	.0000	.0000	.0001	.0002

r \ p	.11	.12	.13	.14	.15	.16	.17	.18	.19	.20
1	.8044	.8330	.8577	.8789	.8972	.9129	.9264	.9379	.9477	.9560
2	.4658	.5141	.5599	.6031	.6433	.6807	.7152	.7469	.7758	.8021
3	.1938	.2315	.2708	.3111	.3521	.3932	.4341	.4744	.5138	.5519
4	.0594	.0774	.0979	.1210	.1465	.1742	.2038	.2351	.2679	.3018
5	.0137	.0196	.0269	.0359	.0467	.0594	.0741	.0907	.1093	.1298
6	.0024	.0038	.0057	.0082	.0115	.0157	.0209	.0273	.0349	.0439
7	.0003	.0006	.0009	.0015	.0022	.0032	.0046	.0064	.0087	.0116
8	.0000	.0001	.0001	.0002	.0003	.0005	.0008	.0012	.0017	.0024
9	.0000	.0000	.0000	.0000	.0000	.0001	.0001	.0002	.0003	.0004

r \ p	.21	.22	.23	.24	.25	.26	.27	.28	.29	.30
1	.9631	.9691	.9742	.9786	.9822	.9852	.9878	.9899	.9917	.9932
2	.8259	.8473	.8665	.8837	.8990	.9126	.9246	.9352	.9444	.9525
3	.5887	.6239	.6574	.6891	.7189	.7467	.7727	.7967	.8188	.8392
4	.3366	.3719	.4076	.4432	.4787	.5136	.5479	.5813	.6137	.6448
5	.1523	.1765	.2023	.2297	.2585	.2884	.3193	.3509	.3832	.4158
6	.0543	.0662	.0797	.0949	.1117	.1301	.1502	.1718	.1949	.2195
7	.0152	.0196	.0248	.0310	.0383	.0467	.0563	.0673	.0796	.0933
8	.0033	.0045	.0060	.0079	.0103	.0132	.0167	.0208	.0257	.0315
9	.0006	.0008	.0011	.0016	.0022	.0029	.0038	.0050	.0065	.0083
10	.0001	.0001	.0002	.0002	.0003	.0005	.0007	.0009	.0012	.0017
11	.0000	.0000	.0000	.0000	.0000	.0001	.0001	.0001	.0002	.0002

r \ p	.31	.32	.33	.34	.35	.36	.37	.38	.39	.40
1	.9945	.9955	.9963	.9970	.9976	.9981	.9984	.9988	.9990	.9992
2	.9596	.9657	.9710	.9756	.9795	.9828	.9857	.9881	.9902	.9919
3	.8577	.8746	.8899	.9037	.9161	.9271	.9370	.9457	.9534	.9602
4	.6747	.7032	.7301	.7556	.7795	.8018	.8226	.8418	.8595	.8757
5	.4486	.4813	.5138	.5458	.5773	.6080	.6378	.6666	.6943	.7207
6	.2454	.2724	.3006	.3297	.3595	.3899	.4208	.4519	.4831	.5141
7	.1084	.1250	.1431	.1626	.1836	.2059	.2296	.2545	.2805	.3075
8	.0381	.0458	.0545	.0643	.0753	.0876	.1012	.1162	.1325	.1501
9	.0105	.0131	.0163	.0200	.0243	.0294	.0353	.0420	.0497	.0583
10	.0022	.0029	.0037	.0048	.0060	.0076	.0095	.0117	.0144	.0175

Table A.1—*Continued*

n = 14 (Continued)

r \ p	.31	.32	.33	.34	.35	.36	.37	.38	.39	.40
11	.0003	.0005	.0006	.0008	.0011	.0014	.0019	.0024	.0031	.0039
12	.0000	.0001	.0001	.0001	.0001	.0002	.0003	.0003	.0005	.0006
13	.0000	.0000	.0000	.0000	.0000	.0000	.0000	.0000	.0000	.0001
	.41	.42	.43	.44	.45	.46	.47	.48	.49	.50
1	.9994	.9995	.9996	.9997	.9998	.9998	.9999	.9999	.9999	.9999
2	.9934	.9946	.9956	.9964	.9971	.9977	.9981	.9985	.9988	.9991
3	.9661	.9713	.9758	.9797	.9830	.9858	.9883	.9903	.9921	.9935
4	.8905	.9039	.9161	.9270	.9368	.9455	.9532	.9601	.9661	.9713
5	.7459	.7697	.7922	.8132	.8328	.8510	.8678	.8833	.8974	.9102
6	.5450	.5754	.6052	.6344	.6627	.6900	.7163	.7415	.7654	.7880
7	.3355	.3643	.3937	.4236	.4539	.4843	.5148	.5451	.5751	.6047
8	.1692	.1896	.2113	.2344	.2586	.2840	.3105	.3380	.3663	.3953
9	.0680	.0789	.0910	.1043	.1189	.1348	.1520	.1707	.1906	.2120
10	.0212	.0255	.0304	.0361	.0426	.0500	.0583	.0677	.0782	.0898
11	.0049	.0061	.0076	.0093	.0114	.0139	.0168	.0202	.0241	.0287
12	.0008	.0010	.0013	.0017	.0022	.0027	.0034	.0042	.0053	.0065
13	.0001	.0001	.0001	.0002	.0003	.0003	.0004	.0006	.0007	.0009
14	.0000	.0000	.0000	.0000	.0000	.0000	.0000	.0000	.0000	.0001

n = 15

r \ p	.01	.02	.03	.04	.05	.06	.07	.08	.09	.10
1	.1399	.2614	.3667	.4579	.5367	.6047	.6633	.7137	.7570	.7941
2	.0096	.0353	.0730	.1191	.1710	.2262	.2832	.3403	.3965	.4510
3	.0004	.0030	.0094	.0203	.0362	.0571	.0829	.1130	.1469	.1841
4	.0000	.0002	.0008	.0024	.0055	.0104	.0175	.0273	.0399	.0556
5	.0000	.0000	.0001	.0002	.0006	.0014	.0028	.0050	.0082	.0127
6	.0000	.0000	.0000	.0000	.0001	.0001	.0003	.0007	.0013	.0022
7	.0000	.0000	.0000	.0000	.0000	.0000	.0000	.0001	.0002	.0003
	.11	.12	.13	.14	.15	.16	.17	.18	.19	.20
1	.8259	.8530	.8762	.8959	.9126	.9269	.9389	.9490	.9576	.9648
2	.5031	.5524	.5987	.6417	.6814	.7179	.7511	.7813	.8085	.8329
3	.2238	.2654	.3084	.3520	.3958	.4392	.4819	.5234	.5635	.6020
4	.0742	.0959	.1204	.1476	.1773	.2092	.2429	.2782	.3146	.3518
5	.0187	.0265	.0361	.0478	.0617	.0778	.0961	.1167	.1394	.1642
6	.0037	.0057	.0084	.0121	.0168	.0227	.0300	.0387	.0490	.0611
7	.0006	.0010	.0015	.0024	.0036	.0052	.0074	.0102	.0137	.0181
8	.0001	.0001	.0002	.0004	.0006	.0010	.0014	.0021	.0030	.0042
9	.0000	.0000	.0000	.0000	.0001	.0001	.0002	.0003	.0005	.0008
10	.0000	.0000	.0000	.0000	.0000	.0000	.0000	.0000	.0001	.0001
	.21	.22	.23	.24	.25	.26	.27	.28	.29	.30
1	.9709	.9759	.9802	.9837	.9866	.9891	.9911	.9928	.9941	.9953
2	.8547	.8741	.8913	.9065	.9198	.9315	.9417	.9505	.9581	.9647
3	.6385	.6731	.7055	.7358	.7639	.7899	.8137	.8355	.8553	.8732
4	.3895	.4274	.4650	.5022	.5387	.5742	.6086	.6416	.6732	.7031
5	.1910	.2195	.2495	.2810	.3135	.3469	.3810	.4154	.4500	.4845
6	.0748	.0905	.1079	.1272	.1484	.1713	.1958	.2220	.2495	.2784
7	.0234	.0298	.0374	.0463	.0566	.0684	.0817	.0965	.1130	.1311
8	.0058	.0078	.0104	.0135	.0173	.0219	.0274	.0338	.0413	.0500
9	.0011	.0016	.0023	.0031	.0042	.0056	.0073	.0094	.0121	.0152
10	.0002	.0003	.0004	.0006	.0008	.0011	.0015	.0021	.0028	.0037
11	.0000	.0000	.0001	.0001	.0001	.0002	.0002	.0003	.0005	.0007
12	.0000	.0000	.0000	.0000	.0000	.0000	.0000	.0000	.0001	.0001
	.31	.32	.33	.34	.35	.36	.37	.38	.39	.40
1	.9962	.9969	.9975	.9980	.9984	.9988	.9990	.9992	.9994	.9995
2	.9704	.9752	.9794	.9829	.9858	.9883	.9904	.9922	.9936	.9948
3	.8893	.9038	.9167	.9281	.9383	.9472	.9550	.9618	.9678	.9729
4	.7314	.7580	.7829	.8060	.8273	.8469	.8649	.8813	.8961	.9095
5	.5187	.5523	.5852	.6171	.6481	.6778	.7062	.7332	.7587	.7827
6	.3084	.3393	.3709	.4032	.4357	.4684	.5011	.5335	.5654	.5968
7	.1509	.1722	.1951	.2194	.2452	.2722	.3003	.3295	.3595	.3902
8	.0599	.0711	.0837	.0977	.1132	.1302	.1487	.1687	.1902	.2131
9	.0190	.0236	.0289	.0351	.0422	.0504	.0597	.0702	.0820	.0950
10	.0048	.0062	.0079	.0099	.0124	.0154	.0190	.0232	.0281	.0338
11	.0009	.0012	.0016	.0022	.0028	.0037	.0047	.0059	.0075	.0093
12	.0001	.0002	.0003	.0004	.0005	.0006	.0009	.0011	.0015	.0019
13	.0000	.0000	.0000	.0000	.0001	.0001	.0001	.0002	.0002	.0003

Table A.1—*Continued*

n = 15 (Continued)

r \ p	.41	.42	.43	.44	.45	.46	.47	.48	.49	.50
1	.9996	.9997	.9998	.9998	.9999	.9999	.9999	.9999	1.0000	1.0000
2	.9958	.9966	.9973	.9979	.9983	.9987	.9990	.9992	.9994	.9995
3	.9773	.9811	.9843	.9870	.9893	.9913	.9929	.9943	.9954	.9963
4	.9215	.9322	.9417	.9502	.9576	.9641	.9697	.9746	.9788	.9824
5	.8052	.8261	.8454	.8633	.8796	.8945	.9080	.9201	.9310	.9408
6	.6274	.6570	.6856	.7131	.7392	.7641	.7875	.8095	.8301	.8491
7	.4214	.4530	.4847	.5164	.5478	.5789	.6095	.6394	.6684	.6964
8	.2374	.2630	.2898	.3176	.3465	.3762	.4065	.4374	.4686	.5000
9	.1095	.1254	.1427	.1615	.1818	.2034	.2265	.2510	.2767	.3036
10	.0404	.0479	.0565	.0661	.0769	.0890	.1024	.1171	.1333	.1509
11	.0116	.0143	.0174	.0211	.0255	.0305	.0363	.0430	.0506	.0592
12	.0025	.0032	.0040	.0051	.0063	.0079	.0097	.0119	.0145	.0176
13	.0004	.0005	.0007	.0009	.0011	.00p4	.0018	.0023	.0029	.0037
14	.0000	.0000	.0001	.0001	.0001	.0002	.0002	.0003	.0004	.0005

n = 16

r \ p	.01	.02	.03	.04	.05	.06	.07	.08	.09	.10
1	.1485	.2762	.3857	.4796	.5599	.6284	.6869	.7366	.7789	.8147
2	.0109	.0399	.0818	.1327	.1892	.2489	.3098	.3701	.4289	.4853
3	.0005	.0037	.0113	.0242	.0429	.0673	.0969	.1311	.1694	.2108
4	.0000	.0002	.0011	.0032	.0070	.0132	.0221	.0342	.0496	.0684
5	.0000	.0000	.0001	.0003	.0009	.0019	.0038	.0068	.0111	.0170
6	.0000	.0000	.0000	.0000	.0001	.0002	.0005	.0010	.0019	.0033
7	.0000	.0000	.0000	.0000	.0000	.0000	.0001	.0001	.0003	.0005
8	.0000	.0000	.0000	.0000	.0000	.0000	.0000	.0000	.0000	.0001
	.11	.12	.13	.14	.15	.16	.17	.18	.19	.20
1	.8450	.8707	.8923	.9105	.9257	.9386	.9493	.9582	.9657	.9719
2	.5386	.5885	.6347	.6773	.7161	.7513	.7830	.8115	.8368	.8593
3	.2545	.2999	.3461	.3926	.4386	.4838	.5277	.5698	.6101	.6482
4	.0907	.1162	.1448	.1763	.2101	.2460	.2836	.3223	.3619	.4019
5	.0248	.0348	.0471	.0618	.0791	.0988	.1211	.1458	.1727	.2018
6	.0053	.0082	.0120	.0171	.0235	.0315	.0412	.0527	.0662	.0817
7	.0009	.0015	.0024	.0038	.0056	.0080	.0112	.0153	.0204	.0267
8	.0001	.0002	.0004	.0007	.0011	.0016	.0024	.0036	.0051	.0070
9	.0000	.0000	.0001	.0001	.0002	.0003	.0004	.0007	.0010	.0015
10	.0000	.0000	.0000	.0000	.0000	.0000	.0001	.0001	.0002	.0002
	.21	.22	.23	.24	.25	.26	.27	.28	.29	.30
1	.9770	.9812	.9847	.9876	.9900	.9919	.9935	.9948	.9958	.9967
2	.8791	.8965	.9117	.9250	.9365	.9465	.9550	.9623	.9686	.9739
3	.6839	.7173	.7483	.7768	.8029	.8267	.8482	.8677	.8851	.9006
4	.4418	.4814	.5203	.5583	.5950	.6303	.6640	.6959	.7260	.7541
5	.2327	.2652	.2991	.3341	.3698	.4060	.4425	.4788	.5147	.5501
6	.0992	.1188	.1405	.1641	.1897	.2169	.2458	.2761	.3077	.3402
7	.0342	.0432	.0536	.0658	.0796	.0951	.1125	.1317	.1526	.1753
8	.0095	.0127	.0166	.0214	.0271	.0340	.0420	.0514	.0621	.0744
9	.0021	.0030	.0041	.0056	.0075	.0098	.0127	.0163	.0206	.0257
10	.0004	.0006	.0008	.0012	.0016	.0023	.0031	.0041	.0055	.0071
11	.0001	.0001	.0001	.0002	.0003	.0004	.0006	.0008	.0011	.0016
12	.0000	.0000	.0000	.0000	.0000	.0001	.0001	.0001	.0002	.0003
	.31	.32	.33	.34	.35	.36	.37	.38	.39	.40
1	.9974	.9979	.9984	.9987	.9990	.9992	.9994	.9995	.9996	.9997
2	.9784	.9822	.9854	.9880	.9902	.9921	.9936	.9948	.9959	.9967
3	.9144	.9266	.9374	.9467	.9549	.9620	.9681	.9734	.9778	.9817
4	.7804	.8047	.8270	.8475	.8661	.8830	.8982	.9119	.9241	.9349
5	.5846	.6181	.6504	.6813	.7108	.7387	.7649	.7895	.8123	.8334
6	.3736	.4074	.4416	.4759	.5100	.5438	.5770	.6094	.6408	.6712
7	.1997	.2257	.2531	.2819	.3119	.3428	.3746	.4070	.4398	.4728
8	.0881	.1035	.1205	.1391	.1594	.1813	.2048	.2298	.2562	.2839
9	.0317	.0388	.0470	.0564	.0671	.0791	.0926	.1076	.1242	.1423
10	.0092	.0117	.0148	.0185	.0229	.0280	.0341	.0411	.0491	.0583
11	.0021	.0028	.0037	.0048	.0062	.0079	.0100	.0125	.0155	.0191
12	.0004	.0005	.0007	.0010	.0013	.0017	.0023	.0030	.0038	.0049
13	.0001	.0001	.0001	.0001	.0002	.0003	.0004	.0005	.0007	.0009
14	.0000	.0000	.0000	.0000	.0000	.0000	.0000	.0001	.0001	.0001

Table A.1—*Continued*

n = 16 (Continued)

r \ p	.41	.42	.43	.44	.45	.46	.47	.48	.49	.50
1	.9998	.9998	.9999	.9999	.9999	.9999	1.0000	1.0000	1.0000	1.0000
2	.9974	.9979	.9984	.9987	.9990	.9992	.9994	.9995	.9997	.9997
3	.9849	.9876	.9899	.9918	.9934	.9947	.9958	.9966	.9973	.9979
4	.9444	.9527	.9600	.9664	.9719	.9766	.9806	.9840	.9869	.9894
5	.8529	.8707	.8869	.9015	.9147	.9265	.9370	.9463	.9544	.9616
6	.7003	.7280	.7543	.7792	.8024	.8241	.8441	.8626	.8795	.8949
7	.5058	.5387	.5711	.6029	.6340	.6641	.6932	.7210	.7476	.7728
8	.3128	.3428	.3736	.4051	.4371	.4694	.5019	.5343	.5665	.5982
9	.1619	.1832	.2060	.2302	.2559	.2829	.3111	.3405	.3707	.4018
10	.0687	.0805	.0936	.1081	.1241	.1416	.1607	.1814	.2036	.2272
11	.0234	.0284	.0342	.0409	.0486	.0574	.0674	.0786	.0911	.1051
12	.0062	.0078	.0098	.0121	.0149	.0183	.0222	.0268	.0322	.0384
13	.0012	.0016	.0021	.0027	.0035	.0044	.0055	.0069	.0086	.0106
14	.0002	.0002	.0003	.0004	.0006	.0007	.0010	.0013	.0016	.0021
15	.0000	.0000	.0000	.0000	.0001	.0001	.0001	.0001	.0002	.0003

n = 17

r \ p	.01	.02	.03	.04	.05	.06	.07	.08	.09	.10
1	.1571	.2907	.4042	.5004	.5819	.6507	.7088	.7577	.7988	.8332
2	.0123	.0446	.0909	.1465	.2078	.2717	.3362	.3995	.4604	.5182
3	.0006	.0044	.0134	.0286	.0503	.0782	.1118	.1503	.1927	.2382
4	.0000	.0003	.0014	.0040	.0088	.0164	.0273	.0419	.0603	.0826
5	.0000	.0000	.0001	.0004	.0012	.0026	.0051	.0090	.0145	.0221
6	.0000	.0000	.0000	.0000	.0001	.0003	.0007	.0015	.0027	.0047
7	.0000	.0000	.0000	.0000	.0000	.0000	.0001	.0002	.0004	.0008
8	.0000	.0000	.0000	.0000	.0000	.0000	.0000	.0000	.0000	.0001

	.11	.12	.13	.14	.15	.16	.17	.18	.19	.20
1	.8621	.8862	.9063	.9230	.9369	.9484	.9579	.9657	.9722	.9775
2	.5723	.6223	.6682	.7099	.7475	.7813	.8113	.8379	.8613	.8818
3	.2858	.3345	.3836	.4324	.4802	.5266	.5711	.6133	.6532	.6904
4	.1087	.1383	.1710	.2065	.2444	.2841	.3251	.3669	.4091	.4511
5	.0321	.0446	.0598	.0778	.0987	.1224	.1487	.1775	.2087	.2418
6	.0075	.0114	.0166	.0234	.0319	.0423	.0548	.0695	.0864	.1057
7	.0014	.0023	.0037	.0056	.0083	.0118	.0163	.0220	.0291	.0377
8	.0002	.0004	.0007	.0011	.0017	.0027	.0039	.0057	.0080	.0109
9	.0000	.0001	.0001	.0002	.0003	.0005	.0008	.0012	.0018	.0026
10	.0000	.0000	.0000	.0000	.0000	.0001	.0001	.0002	.0003	.0005
11	.0000	.0000	.0000	.0000	.0000	.0000	.0000	.0000	.0000	.0001

	.21	.22	.23	.24	.25	.26	.27	.28	.29	.30
1	.9818	.9854	.9882	.9906	.9925	.9940	.9953	.9962	.9970	.9977
2	.8996	.9152	.9285	.9400	.9499	.9583	.9654	.9714	.9765	.9807
3	.7249	.7567	.7859	.8123	.8363	.8578	.8771	.8942	.9093	.9226
4	.4927	.5333	.5728	.6107	.6470	.6814	.7137	.7440	.7721	.7981
5	.2766	.3128	.3500	.3879	.4261	.4643	.5023	.5396	.5760	.6113
6	.1273	.1510	.1770	.2049	.2347	.2661	.2989	.3329	.3677	.4032
7	.0479	.0598	.0736	.0894	.1071	.1268	.1485	.1721	.1976	.2248
8	.0147	.0194	.0251	.0320	.0402	.0499	.0611	.0739	.0884	.1046
9	.0037	.0051	.0070	.0094	.0124	.0161	.0206	.0261	.0326	.0403
10	.0007	.0011	.0016	.0022	.0031	.0042	.0057	.0075	.0098	.0127
11	.0001	.0002	.0003	.0004	.0006	.0009	.0013	.0018	.0024	.0032
12	.0000	.0000	.0000	.0001	.0001	.0002	.0002	.0003	.0005	.0007
13	.0000	.0000	.0000	.0000	.0000	.0000	.0000	.0000	.0001	.0001

	.31	.32	.33	.34	.35	.36	.37	.38	.39	.40
1	.9982	.9986	.9989	.9991	.9993	.9995	.9996	.9997	.9998	.9998
2	.9843	.9872	.9896	.9917	.9933	.9946	.9957	.9966	.9973	.9979
3	.9343	.9444	.9532	.9608	.9673	.9728	.9775	.9815	.9849	.9877
4	.8219	.8437	.8634	.8812	.8972	.9115	.9241	.9353	.9450	.9536
5	.6453	.6778	.7087	.7378	.7652	.7906	.8142	.8360	.8559	.8740
6	.4390	.4749	.5105	.5458	.5803	.6139	.6465	.6778	.7077	.7361
7	.2536	.2838	.3153	.3479	.3812	.4152	.4495	.4839	.5182	.5522
8	.1227	.1426	.1642	.1877	.2128	.2395	.2676	.2971	.3278	.3595
9	.0492	.0595	.0712	.0845	.0994	.1159	.1341	.1541	.1757	.1989
10	.0162	.0204	.0254	.0314	.0383	.0464	.0557	.0664	.0784	.0919

Table A.1—*Continued*

n = 17 (Continued)

r \ p	.31	.32	.33	.34	.35	.36	.37	.38	.39	.40
11	.0043	.0057	.0074	.0095	.0120	.0151	.0189	.0234	.0286	.0348
12	.0009	.0013	.0017	.0023	.0030	.0040	.0051	.0066	.0084	.0106
13	.0002	.0002	.0003	.0004	.0006	.0008	.0011	.0015	.0019	.0025
14	.0000	.0000	.0000	.0001	.0001	.0001	.0002	.0002	.0003	.0005
15	.0000	.0000	.0000	.0000	.0000	.0000	.0000	.0000	.0000	.0001

r \ p	.41	.42	.43	.44	.45	.46	.47	.48	.49	.50
1	.9999	.9999	.9999	.9999	1.0000	1.0000	1.0000	1.0000	1.0000	1.0000
2	.9984	.9987	.9990	.9992	.9994	.9996	.9997	.9998	.9998	.9999
3	.9900	.9920	.9935	.9948	.9959	.9968	.9975	.9980	.9985	.9988
4	.9610	.9674	.9729	.9776	.9816	.9849	.9877	.9901	.9920	.9936
5	.8904	.9051	.9183	.9301	.9404	.9495	.9575	.9644	.9704	.9755
6	.7628	.7879	.8113	.8330	.8529	.8712	.8878	.9028	.9162	.9283
7	.5856	.6182	.6499	.6805	.7098	.7377	.7641	.7890	.8122	.8338
8	.3920	.4250	.4585	.4921	.5257	.5590	.5918	.6239	.6552	.6855
9	.2238	.2502	.2780	.3072	.3374	.3687	.4008	.4335	.4667	.5000
10	.1070	.1236	.1419	.1618	.1834	.2066	.2314	.2577	.2855	.3145
11	.0420	.0503	.0597	.0705	.0826	.0962	.1112	.1279	.1462	.1662
12	.0133	.0165	.0203	.0248	.0301	.0363	.0434	.0517	.0611	.0717
13	.0033	.0042	.0054	.0069	.0086	.0108	.0134	.0165	.0202	.0245
14	.0006	.0008	.0011	.0014	.0019	.0024	.0031	.0040	.0050	.0064
15	.0001	.0001	.0002	.0002	.0003	.0004	.0005	.0007	.0009	.0012
16	.0000	.0000	.0000	.0000	.0000	.0000	.0001	.0001	.0001	.0001

n = 18

r \ p	.01	.02	.03	.04	.05	.06	.07	.08	.09	.10
1	.1655	.3049	.4220	.5204	.6028	.6717	.7292	.7771	.8169	.8499
2	.0138	.0495	.1003	.1607	.2265	.2945	.3622	.4281	.4909	.5497
3	.0007	.0052	.0157	.0333	.0581	.0898	.1275	.1702	.2168	.2662
4	.0000	.0004	.0018	.0050	.0109	.0201	.0333	.0506	.0723	.0982
5	.0000	.0000	.0002	.0006	.0015	.0034	.0067	.0116	.0187	.0282
6	.0000	.0000	.0000	.0001	.0002	.0005	.0010	.0021	.0038	.0064
7	.0000	.0000	.0000	.0000	.0000	.0000	.0001	.0003	.0006	.0012
8	.0000	.0000	.0000	.0000	.0000	.0000	.0000	.0000	.0001	.0002

r \ p	.11	.12	.13	.14	.15	.16	.17	.18	.19	.20
1	.8773	.8998	.9185	.9338	.9464	.9566	.9651	.9719	.9775	.9820
2	.6042	.6540	.6992	.7398	.7759	.8080	.8362	.8609	.8824	.9009
3	.3173	.3690	.4206	.4713	.5203	.5673	.6119	.6538	.6927	.7287
4	.1282	.1618	.1986	.2382	.2798	.3229	.3669	.4112	.4554	.4990
5	.0405	.0558	.0743	.0959	.1206	.1482	.1787	.2116	.2467	.2836
6	.0102	.0154	.0222	.0310	.0419	.0551	.0708	.0889	.1097	.1329
7	.0021	.0034	.0054	.0081	.0118	.0167	.0229	.0306	.0400	.0513
8	.0003	.0006	.0011	.0017	.0027	.0041	.0060	.0086	.0120	.0163
9	.0000	.0001	.0002	.0003	.0005	.0008	.0013	.0020	.0030	.0043
10	.0000	.0000	.0000	.0000	.0001	.0001	.0002	.0004	.0006	.0009
11	.0000	.0000	.0000	.0000	.0000	.0000	.0000	.0001	.0001	.0002

r \ p	.21	.22	.23	.24	.25	.26	.27	.28	.29	.30
1	.9856	.9886	.9909	.9928	.9944	.9956	.9965	.9973	.9979	.9984
2	.9169	.9306	.9423	.9522	.9605	.9676	.9735	.9784	.9824	.9858
3	.7616	.7916	.8187	.8430	.8647	.8839	.9009	.9158	.9288	.9400
4	.5414	.5825	.6218	.6591	.6943	.7272	.7578	.7860	.8119	.8354
5	.3220	.3613	.4012	.4414	.4813	.5208	.5594	.5968	.6329	.6673
6	.1586	.1866	.2168	.2488	.2825	.3176	.3538	.3907	.4281	.4656
7	.0645	.0799	.0974	.1171	.1390	.1630	.1891	.2171	.2469	.2783
8	.0217	.0283	.0363	.0458	.0569	.0699	.0847	.1014	.1200	.1407
9	.0060	.0083	.0112	.0148	.0193	.0249	.0316	.0395	.0488	.0596
10	.0014	.0020	.0028	.0039	.0054	.0073	.0097	.0127	.0164	.0210
11	.0003	.0004	.0006	.0009	.0012	.0018	.0025	.0034	.0046	.0061
12	.0000	.0001	.0001	.0002	.0002	.0003	.0005	.0007	.0010	.0014
13	.0000	.0000	.0000	.0000	.0000	.0001	.0001	.0001	.0002	.0003

r \ p	.31	.32	.33	.34	.35	.36	.37	.38	.39	.40
1	.9987	.9990	.9993	.9994	.9996	.9997	.9998	.9998	.9999	.9999
2	.9886	.9908	.9927	.9942	.9954	.9964	.9972	.9978	.9983	.9987
3	.9498	.9581	.9652	.9713	.9764	.9807	.9843	.9873	.9897	.9918
4	.8568	.8759	.8931	.9083	.9217	.9335	.9439	.9528	.9606	.9672
5	.7001	.7309	.7598	.7866	.8114	.8341	.8549	.8737	.8907	.9058

Table A.1—*Continued*

n = 18 (Continued)

r \ p	.31	.32	.33	.34	.35	.36	.37	.38	.39	.40
6	.5029	.5398	.5759	.6111	.6450	.6776	.7086	.7379	.7655	.7912
7	.3111	.3450	.3797	.4151	.4509	.4867	.5224	.5576	.5921	.6257
8	.1633	.1878	.2141	.2421	.2717	.3027	.3349	.3681	.4021	.4366
9	.0720	.0861	.1019	.1196	.1391	.1604	.1835	.2084	.2350	.2632
10	.0264	.0329	.0405	.0494	.0597	.0714	.0847	.0997	.1163	.1347
11	.0080	.0104	.0133	.0169	.0212	.0264	.0325	.0397	.0480	.0576
12	.0020	.0027	.0036	.0047	.0062	.0080	.0102	.0130	.0163	.0203
13	.0004	.0005	.0008	.0011	.0014	.0019	.0026	.0034	.0044	.0058
14	.0001	.0001	.0001	.0002	.0003	.0004	.0005	.0007	.0010	.0013
15	.0000	.0000	.0000	.0000	.0000	.0001	.0001	.0001	.0002	.0002

r \ p	.41	.42	.43	.44	.45	.46	.47	.48	.49	.50
1	.9999	.9999	1.0000	1.0000	1.0000	1.0000	1.0000	1.0000	1.0000	1.0000
2	.9990	.9992	.9994	.9996	.9997	.9998	.9998	.9999	.9999	.9999
3	.9934	.9948	.9959	.9968	.9975	.9981	.9985	.9989	.9991	.9993
4	.9729	.9777	.9818	.9852	.9880	.9904	.9923	.9939	.9952	.9962
5	.9193	.9313	.9418	.9510	.9589	.9658	.9717	.9767	.9810	.9846
6	.8151	.8372	.8573	.8757	.8923	.9072	.9205	.9324	.9428	.9519
7	.6582	.6895	.7193	.7476	.7742	.7991	.8222	.8436	.8632	.8811
8	.4713	.5062	.5408	.5750	.6085	.6412	.6728	.7032	.7322	.7597
9	.2928	.3236	.3556	.3885	.4222	.4562	.4906	.5249	.5591	.5927
10	.1549	.1768	.2004	.2258	.2527	.2812	.3110	.3421	.3742	.4073
11	.0686	.0811	.0951	.1107	.1280	.1470	.1677	.1902	.2144	.2403
12	.0250	.0307	.0372	.0449	.0537	.0638	.0753	.0883	.1028	.1189
13	.0074	.0094	.0118	.0147	.0183	.0225	.0275	.0334	.0402	.0481
14	.0017	.0022	.0029	.0038	.0049	.0063	.0079	.0100	.0125	.0154
15	.0003	.0004	.0006	.0007	.0010	.0013	.0017	.0023	.0029	.0038
16	.0000	.0001	.0001	.0001	.0001	.0002	.0003	.0004	.0005	.0007
17	.0000	.0000	.0000	.0000	.0000	.0000	.0000	.0000	.0001	.0001

n = 19

r \ p	.01	.02	.03	.04	.05	.06	.07	.08	.09	.10
1	.1738	.3188	.4394	.5396	.6226	.6914	.7481	.7949	.8334	.8649
2	.0153	.0546	.1100	.1751	.2453	.3171	.3879	.4560	.5202	.5797
3	.0009	.0061	.0083	.0384	.0665	.1021	.1439	.1908	.2415	.2946
4	.0000	.0005	.0002	.0061	.0132	.0243	.0398	.0602	.0853	.1150
5	.0000	.0000	.0002	.0007	.0020	.0044	.0085	.0147	.0235	.0352
6	.0000	.0000	.0000	.0001	.0002	.0006	.0014	.0029	.0051	.0086
7	.0000	.0000	.0000	.0000	.0000	.0001	.0002	.0004	.0009	.0017
8	.0000	.0000	.0000	.0000	.0000	.0000	.0000	.0001	.0001	.0003

r \ p	.11	.12	.13	.14	.15	.16	.17	.18	.19	.20
1	.8908	.9119	.9291	.9431	.9544	.9636	.9710	.9770	.9818	.9856
2	.6342	.6835	.7277	.7669	.8015	.8318	.8581	.8809	.9004	.9171
3	.3488	.4032	.4568	.5089	.5587	.6059	.6500	.6910	.7287	.7631
4	.1490	.1867	.2275	.2708	.3159	.3620	.4085	.4549	.5005	.5449
5	.0502	.0685	.0904	.1158	.1444	.1762	.2107	.2476	.2864	.3267
6	.0135	.0202	.0290	.0401	.0537	.0700	.0891	.1110	.1357	.1631
7	.0030	.0048	.0076	.0113	.0163	.0228	.0310	.0411	.0532	.0676
8	.0005	.0009	.0016	.0026	.0041	.0061	.0089	.0126	.0173	.0233
9	.0001	.0002	.0003	.0005	.0008	.0014	.0021	.0032	.0047	.0067
10	.0000	.0000	.0000	.0001	.0001	.0002	.0004	.0007	.0010	.0016
11	.0000	.0000	.0000	.0000	.0000	.0000	.0001	.0001	.0002	.0003
12	.0000	.0000	.0000	.0000	.0000	.0000	.0000	.0000	.0000	.0001

r \ p	.21	.22	.23	.24	.25	.26	.27	.28	.29	.30
1	.9887	.9911	.9930	.9946	.9958	.9967	.9975	.9981	.9985	.9989
2	.9313	.9434	.9535	.9619	.9690	.9749	.9797	.9837	.9869	.9896
3	.7942	.8222	.8471	.8692	.8887	.9057	.9205	.9333	.9443	.9538
4	.5877	.6285	.6671	.7032	.7369	.7680	.7965	.8224	.8458	.8668
5	.3681	.4100	.4520	.4936	.5346	.5744	.6129	.6498	.6848	.7178
6	.1929	.2251	.2592	.2950	.3322	.3705	.4093	.4484	.4875	.5261
7	.0843	.1034	.1249	.1487	.1749	.2032	.2336	.2657	.2995	.3345
8	.0307	.0396	.0503	.0629	.0775	.0941	.1129	.1338	.1568	.1820
9	.0093	.0127	.0169	.0222	.0287	.0366	.0459	.0568	.0694	.0839
10	.0023	.0034	.0047	.0066	.0089	.0119	.0156	.0202	.0258	.0326

Table A.1—*Continued*

n = 19 (Continued)

r \ p	.21	.22	.23	.24	.25	.26	.27	.28	.29	.30
11	.0005	.0007	.0011	.0016	.0023	.0032	.0044	.0060	.0080	.0105
12	.0001	.0001	.0002	.0003	.0005	.0007	.0010	.0015	.0021	.0028
13	.0000	.0000	.0000	.0001	.0001	.0001	.0002	.0003	.0004	.0006
14	.0000	.0000	.0000	.0000	.0000	.0000	.0000	.0000	.0001	.0001

r \ p	.31	.32	.33	.34	.35	.36	.37	.38	.39	.40
1	.9991	.9993	.9995	.9996	.9997	.9998	.9998	.9999	.9999	.9999
2	.9917	.9935	.9949	.9960	.9969	.9976	.9981	.9986	.9989	.9992
3	.9618	.9686	.9743	.9791	.9830	.9863	.9890	.9913	.9931	.9945
4	.8856	.9022	.9169	.9297	.9409	.9505	.9588	.9659	.9719	.9770
5	.7486	.7773	.8037	.8280	.8500	.8699	.8878	.9038	.9179	.9304
6	.5641	.6010	.6366	.6707	.7032	.7339	.7627	.7895	.8143	.8371
7	.3705	.4073	.4445	.4818	.5188	.5554	.5913	.6261	.6597	.6919
8	.2091	.2381	.2688	.3010	.3344	.3690	.4043	.4401	.4762	.5122
9	.1003	.1186	.1389	.1612	.1855	.2116	.2395	.2691	.3002	.3325
10	.0405	.0499	.0608	.0733	.0875	.1035	.1213	.1410	.1626	.1861
11	.0137	.0176	.0223	.0280	.0347	.0426	.0518	.0625	.0747	.0885
12	.0038	.0051	.0068	.0089	.0114	.0146	.0185	.0231	.0287	.0352
13	.0009	.0012	.0017	.0023	.0031	.0041	.0054	.0070	.0091	.0116
14	.0002	.0002	.0003	.0005	.0007	.0009	.0013	.0017	.0023	.0031
15	.0000	.0000	.0001	.0001	.0001	.0002	.0002	.0003	.0005	.0006
16	.0000	.0000	.0000	.0000	.0000	.0000	.0000	.0000	.0001	.0001

r \ p	.41	.42	.43	.44	.45	.46	.47	.48	.49	.50
1	1.0000	1.0000	1.0000	1.0000	1.0000	1.0000	1.0000	1.0000	1.0000	1.0000
2	.9994	.9995	.9996	.9997	.9998	.9999	.9999	.9999	.9999	1.0000
3	.9957	.9967	.9974	.9980	.9985	.9988	.9991	.9993	.9995	.9996
4	.9813	.9849	.9878	.9903	.9923	.9939	.9952	.9963	.9971	.9978
5	.9413	.9518	.9590	.9660	.9720	.9771	.9814	.9850	.9879	.9904
6	.8579	.8767	.8937	.9088	.9223	.9342	.9446	.9537	.9615	.9682
7	.7226	.7515	.7787	.8039	.8273	.8488	.8684	.8862	.9022	.9165
8	.5480	.5832	.6176	.6509	.6831	.7138	.7430	.7706	.7964	.8204
9	.3660	.4003	.4353	.4706	.5060	.5413	.5762	.6105	.6439	.6762
10	.2114	.2385	.2672	.2974	.3290	.3617	.3954	.4299	.4648	.5000
11	.1040	.1213	.1404	.1613	.1841	.2087	.2351	.2631	.2928	.3238
12	.0429	.0518	.0621	.0738	.0871	.1021	.1187	.1372	.1575	.1796
13	.0146	.0183	.0227	.0280	.0342	.0415	.0500	.0597	.0709	.0835
14	.0040	.0052	.0068	.0086	.0109	.0137	.0171	.0212	.0261	.0318
15	.0009	.0012	.0016	.0021	.0028	.0036	.0046	.0060	.0076	.0096
16	.0001	.0002	.0003	.0004	.0005	.0007	.0010	.0013	.0017	.0022
17	.0000	.0000	.0000	.0001	.0001	.0001	.0001	.0002	.0003	.0004

n = 20

r \ p	.01	.02	.03	.04	.05	.06	.07	.08	.09	.10
1	.1821	.3324	.4562	.5580	.6415	.7099	.7658	.8113	.8484	.8784
2	.0169	.0599	.1198	.1897	.2642	.3395	.4131	.4831	.5484	.6083
3	.0010	.0071	.0210	.0439	.0755	.1150	.1610	.2121	.2666	.3231
4	.0000	.0006	.0027	.0074	.0159	.0290	.0471	.0706	.0993	.1330
5	.0000	.0000	.0003	.0010	.0026	.0056	.0107	.0183	.0290	.0432
6	.0000	.0000	.0000	.0001	.0003	.0009	.0019	.0038	.0068	.0113
7	.0000	.0000	.0000	.0000	.0000	.0001	.0003	.0006	.0013	.0024
8	.0000	.0000	.0000	.0000	.0000	.0000	.0000	.0001	.0002	.0004
9	.0000	.0000	.0000	.0000	.0000	.0000	.0000	.0000	.0000	.0001

r \ p	.11	.12	.13	.14	.15	.16	.17	.18	.19	.20
1	.9028	.9224	.9383	.9510	.9612	.9694	.9759	.9811	.9852	.9885
2	.6624	.7109	.7539	.7916	.8244	.8529	.8773	.8982	.9159	.9308
3	.3802	.4369	.4920	.5450	.5951	.6420	.6854	.7252	.7614	.7939
4	.1710	.2127	.2573	.3041	.3523	.4010	.4496	.4974	.5439	.5886
5	.0610	.0827	.1083	.1375	.1702	.2059	.2443	.2849	.3271	.3704
6	.0175	.0260	.0370	.0507	.0673	.0870	.1098	.1356	.1643	.1958
7	.0041	.0067	.0103	.0153	.0219	.0304	.0409	.0537	.0689	.0867
8	.0008	.0014	.0024	.0038	.0059	.0088	.0127	.0177	.0241	.0321
9	.0001	.0002	.0005	.0008	.0013	.0021	.0033	.0049	.0071	.0100
10	.0000	.0000	.0001	.0001	.0002	.0004	.0007	.0011	.0017	.0026
11	.0000	.0000	.0000	.0000	.0000	.0001	.0001	.0002	.0004	.0006
12	.0000	.0000	.0000	.0000	.0000	.0000	.0000	.0000	.0001	.0001

Table A.1—*Continued*

n = 20 (Continued)

r \ p	.21	.22	.23	.24	.25	.26	.27	.28	.29	.30
1	.9910	.9931	.9946	.9959	.9968	.9976	.9982	.9986	.9989	.9992
2	.9434	.9539	.9626	.9698	.9757	.9805	.9845	.9877	.9903	.9924
3	.8230	.8488	.8716	.8915	.9087	.9237	.9365	.9474	.9567	.9645
4	.6310	.6711	.7085	.7431	.7748	.8038	.8300	.8534	.8744	.8929
5	.4142	.4580	.5014	.5439	.5852	.6248	.6625	.6981	.7315	.7625
6	.2297	.2657	.3035	.3427	.3828	.4235	.4643	.5048	.5447	.5836
7	.1071	.1301	.1558	.1838	.2142	.2467	.2810	.3169	.3540	.3920
8	.0419	.0536	.0675	.0835	.1018	.1225	.1455	.1707	.1982	.2277
9	.0138	.0186	.0246	.0320	.0409	.0515	.0640	.0784	.0948	.1133
10	.0038	.0054	.0075	.0103	.0139	.0183	.0238	.0305	.0385	.0480
11	.0009	.0013	.0019	.0028	.0039	.0055	.0074	.0100	.0132	.0171
12	.0002	.0003	.0004	.0006	.0009	.0014	.0019	.0027	.0038	.0051
13	.0000	.0000	.0001	.0001	.0002	.0003	.0004	.0006	.0009	.0013
14	.0000	.0000	.0000	.0000	.0000	.0000	.0001	.0001	.0002	.0003

r \ p	.31	.32	.33	.34	.35	.36	.37	.38	.39	.40
1	.9994	.9996	.9997	.9998	.9998	.9999	.9999	.9999	.9999	1.0000
2	.9940	.9953	.9964	.9972	.9979	.9984	.9988	.9991	.9993	.9995
3	.9711	.9765	.9811	.9848	.9879	.9904	.9924	.9940	.9953	.9964
4	.9092	.9235	.9358	.9465	.9556	.9634	.9700	.9755	.9802	.9840
5	.7911	.8173	.8411	.8626	.8818	.8989	.9141	.9274	.9390	.9490
6	.6213	.6574	.6918	.7242	.7546	.7829	.8090	.8329	.8547	.8744
7	.4305	.4693	.5079	.5460	.5834	.6197	.6547	.6882	.7200	.7500
8	.2591	.2922	.3268	.3624	.3990	.4361	.4735	.5108	.5478	.5841
9	.1340	.1568	.1818	.2087	.2376	.2683	.3005	.3341	.3688	.4044
10	.0591	.0719	.0866	.1032	.1218	.1424	.1650	.1897	.2163	.2447
11	.0220	.0279	.0350	.0434	.0532	.0645	.0775	.0923	.1090	.1275
12	.0069	.0091	.0119	.0154	.0196	.0247	.0308	.0381	.0466	.0565
13	.0018	.0025	.0034	.0045	.0060	.0079	.0102	.0132	.0167	.0210
14	.0004	.0006	.0008	.0011	.0015	.0021	.0028	.0037	.0049	.0065
15	.0001	.0001	.0001	.0002	.0003	.0004	.0006	.0009	.0012	.0016
16	.0000	.0000	.0000	.0000	.0001	.0001	.0001	.0002	.0002	.0003

r \ p	.41	.42	.43	.44	.45	.46	.47	.48	.49	.50
1	1.0000	1.0000	1.0000	1.0000	1.0000	1.0000	1.0000	1.0000	1.0000	1.0000
2	.9996	.9997	.9998	.9998	.9999	.9999	.9999	1.0000	1.0000	1.0000
3	.9972	.9979	.9984	.9988	.9991	.9993	.9995	.9996	.9997	.9998
4	.9872	.9898	.9920	.9937	.9951	.9962	.9971	.9977	.9983	.9987
5	.9577	.9651	.9714	.9767	.9811	.9848	.9879	.9904	.9924	.9941
6	.8921	.9078	.9217	.9340	.9447	.9539	.9619	.9687	.9745	.9793
7	.7780	.8041	.8281	.8501	.8701	.8881	.9042	.9186	.9312	.9423
8	.6196	.6539	.6868	.7183	.7480	.7759	.8020	.8261	.8482	.8684
9	.4406	.4771	.5136	.5499	.5847	.6207	.6546	.6873	.7186	.7483
10	.2748	.3064	.3394	.3736	.4086	.4443	.4804	.5166	.5525	.5881
11	.1480	.1705	.1949	.2212	.2493	.2791	.3104	.3432	.3771	.4119
12	.0679	.0810	.0958	.1123	.1308	.1511	.1734	.1977	.2238	.2517
13	.0262	.0324	.0397	.0482	.0580	.0694	.0823	.0969	.1133	.1316
14	.0084	.0107	.0136	.0172	.0214	.0265	.0326	.0397	.0480	.0577
15	.0022	.0029	.0038	.0050	.0064	.0083	.0105	.0133	.0166	.0207
16	.0004	.0006	.0008	.0011	.0015	.0020	.0027	.0035	.0046	.0059
17	.0001	.0001	.0001	.0002	.0003	.0004	.0005	.0007	.0010	.0013
18	.0000	.0000	.0000	.0000	.0000	.0001	.0001	.0001	.0001	.0002

n = 25

r \ p	.01	.02	.03	.04	.05	.06	.07	.08	.09	.10
1	.2222	.3965	.5330	.6396	.7226	.7871	.8370	.8756	.9054	.9282
2	.0258	.0886	.1720	.2642	.3576	.4473	.5304	.6053	.6714	.7288
3	.0020	.0132	.0380	.0765	.1271	.1871	.2534	.3232	.3937	.4629
4	.0001	.0014	.0062	.0165	.0341	.0598	.0936	.1351	.1831	.2364
5	.0000	.0001	.0008	.0028	.0072	.0150	.0274	.0451	.0686	.0980
6	.0000	.0000	.0001	.0004	.0012	.0031	.0065	.0123	.0210	.0334
7	.0000	.0000	.0000	.0000	.0002	.0005	.0013	.0028	.0054	.0095
8	.0000	.0000	.0000	.0000	.0000	.0001	.0002	.0005	.0011	.0023
9	.0000	.0000	.0000	.0000	.0000	.0000	.0000	.0001	.0002	.0005
10	.0000	.0000	.0000	.0000	.0000	.0000	.0000	.0000	.0000	.0001

Table A.1—*Continued*

n = 25 (Continued)

r \ p	.11	.12	.13	.14	.15	.16	.17	.18	.19	.20
1	.9457	.9591	.9692	.9770	.9828	.9872	.9905	.9930	.9948	.9962
2	.7779	.8195	.8543	.8832	.9069	.9263	.9420	.9546	.9646	.9726
3	.5291	.5912	.6483	.7000	.7463	.7870	.8226	.8533	.8796	.9018
4	.2934	.3525	.4123	.4714	.5289	.5837	.6352	.6829	.7266	.7660
5	.1331	.1734	.1283	.2668	.3179	.3707	.4241	.4772	.5292	.5793
6	.0499	.0709	.0965	.1268	.1615	.2002	.2425	.2875	.3347	.3833
7	.0156	.0243	.0359	.0509	.0695	.0920	.1185	.1488	.1827	.2200
8	.0041	.0070	.0113	.0173	.0255	.0361	.0495	.0661	.0859	.1091
9	.0009	.0017	.0030	.0050	.0080	.0121	.0178	.0252	.0348	.0468
10	.0002	.0004	.0007	.0013	.0021	.0035	.0055	.0083	.0122	.0173
11	.0000	.0001	.0001	.0003	.0005	.0009	.0015	.0024	.0037	.0056
12	.0000	.0000	.0000	.0000	.0001	.0002	.0003	.0006	.0010	.0015
13	.0000	.0000	.0000	.0000	.0000	.0000	.0001	.0001	.0002	.0004
14	.0000	.0000	.0000	.0000	.0000	.0000	.0000	.0000	.0000	.0001

r \ p	.21	.22	.23	.24	.25	.26	.27	.28	.29	.30
1	.9972	.9980	.9985	.9990	.9992	.9995	.9996	.9997	.9998	.9999
2	.9789	.9838	.9877	.9907	.9930	.9947	.9961	.9971	.9979	.9984
3	.9204	.9360	.9488	.9593	.9679	.9748	.9804	.9848	.9883	.9910
4	.8013	.8324	.8597	.8834	.9038	.9211	.9358	.9481	.9583	.9668
5	.6270	.6718	.7134	.7516	.7863	.8174	.8452	.8696	.8910	.9095
6	.4325	.4816	.5299	.5767	.6217	.6644	.7044	.7415	.7755	.8065
7	.2601	.3027	.3471	.3927	.4389	.4851	.5308	.5753	.6183	.6593
8	.1358	.1658	.1989	.2349	.2735	.3142	.3565	.3999	.4440	.4882
9	.0614	.0788	.0993	.1228	.1494	.1790	.2115	.2465	.2838	.3231
10	.0240	.0325	.0431	.0560	.0713	.0893	.1101	.1338	.1602	.1894
11	.0082	.0117	.0163	.0222	.0297	.0389	.0502	.0636	.0795	.0978
12	.0024	.0036	.0053	.0076	.0107	.0148	.0199	.0264	.0345	.0442
13	.0006	.0010	.0015	.0023	.0034	.0049	.0069	.0096	.0130	.0175
14	.0001	.0002	.0004	.0006	.0009	.0014	.0021	.0030	.0043	.0060
15	.0000	.0000	.0001	.0001	.0002	.0003	.0005	.0008	.0012	.0018
16	.0000	.0000	.0000	.0000	.0000	.0001	.0001	.0002	.0003	.0005
17	.0000	.0000	.0000	.0000	.0000	.0000	.0000	.0000	.0001	.0001

r \ p	.31	.32	.33	.34	.35	.36	.37	.38	.39	.40
1	.9999	.9999	1.0000	1.0000	1.0000	1.0000	1.0000	1.0000	1.0000	1.0000
2	.9989	.9992	.9994	.9996	.9997	.9998	.9998	.9999	.9999	.9999
3	.9932	.9949	.9961	.9971	.9979	.9984	.9989	.9992	.9994	.9996
4	.9737	.9793	.9838	.9874	.9903	.9926	.9944	.9958	.9968	.9976
5	.9254	.9390	.9504	.9600	.9680	.9745	.9799	.9842	.9877	.9905
6	.8344	.8593	.8813	.9006	.9174	.9318	.9441	.9546	.9633	.9706
7	.6981	.7343	.7679	.7987	.8266	.8517	.8742	.8940	.9114	.9264
8	.5319	.5747	.6163	.6561	.6939	.7295	.7626	.7932	.8211	.8464
9	.3639	.4057	.4482	.4908	.5332	.5748	.6152	.6542	.6914	.7265
10	.2213	.2555	.2919	.3300	.3697	.4104	.4517	.4933	.5347	.5754
11	.1188	.1424	.1686	.1975	.2288	.2624	.2981	.3355	.3743	.4142
12	.0560	.0698	.0859	.1044	.1254	.1490	.1751	.2036	.2346	.2677
13	.0230	.0299	.0383	.0485	.0604	.0745	.0907	.1093	.1303	.1538
14	.0083	.0112	.0149	.0196	.0255	.0326	.0412	.0515	.0637	.0778
15	.0026	.0036	.0050	.0069	.0093	.0124	.0163	.0212	.0271	.0344
16	.0007	.0010	.0015	.0021	.0029	.0041	.0056	.0075	.0100	.0132
17	.0002	.0002	.0004	.0005	.0008	.0011	.0016	.0023	.0032	.0043
18	.0000	.0000	.0001	.0001	.0002	.0003	.0004	.0006	.0008	.0012
19	.0000	.0000	.0000	.0000	.0000	.0001	.0001	.0001	.0002	.0003
20	.0000	.0000	.0000	.0000	.0000	.0000	.0000	.0000	.0000	.0001

r \ p	.41	.42	.43	.44	.45	.46	.47	.48	.49	.50
1	1.0000	1.0000	1.0000	1.0000	1.0000	1.0000	1.0000	1.0000	1.0000	1.0000
2	1.0000	1.0000	1.0000	1.0000	1.0000	1.0000	1.0000	1.0000	1.0000	1.0000
3	.9997	.9998	.9998	.9999	.9999	1.0000	1.0000	1.0000	1.0000	1.0000
4	.9983	.9987	.9991	.9993	.9995	.9997	.9998	.9998	.9999	.9999
5	.9927	.9945	.9958	.9969	.9977	.9983	.9988	.9991	.9994	.9995
6	.9767	.9816	.9856	.9888	.9914	.9934	.9950	.9963	.9972	.9980
7	.9394	.9505	.9599	.9677	.9742	.9796	.9840	.9876	.9904	.9927
8	.8692	.8894	.9071	.9227	.9361	.9477	.9575	.9658	.9727	.9784
9	.7593	.7897	.8177	.8431	.9660	.8865	.9046	.9205	.9343	.9461
10	.6151	.6535	.6902	.7250	.7576	.7880	.8160	.8415	.8646	.8852

Table A.1—*Continued*

n = 25 (Continued)

r \ p	.41	.42	.43	.44	.45	.46	.47	.48	.49	.50
11	.4548	.4956	.5363	.5765	.6157	.6538	.6902	.7249	.7574	.7878
12	.3029	.3397	.3780	.4174	.4574	.4978	.5382	.5780	.6171	.6550
13	.1797	.2080	.2387	.2715	.3063	.3429	.3808	.4199	.4598	.5000
14	.0941	.1127	.1336	.1569	.1827	.2109	.2413	.2740	.3086	.3450
15	.0431	.0535	.0656	.0797	.0960	.1145	.1353	.1585	.1841	.2122
16	.0171	.0220	.0280	.0353	.0440	.0543	.0663	.0803	.0964	.1148
17	.0058	.0078	.0103	.0134	.0174	.0222	.0281	.0352	.0438	.0539
18	.0017	.0023	.0032	.0044	.0058	.0077	.0102	.0132	.0170	.0216
19	.0004	.0006	.0008	.0012	.0016	.0023	.0031	.0041	.0055	.0073
20	.0001	.0001	.0002	.0003	.0004	.0005	.0008	.0011	.0015	.0020
21	.0000	.0000	.0000	.0000	.0001	.0001	.0002	.0002	.0003	.0005
22	.0000	.0000	.0000	.0000	.0000	.0000	.0000	.0000	.0001	.0001

n = 30

r \ p	.01	.02	.03	.04	.05	.06	.07	.08	.09	.10
1	.2603	.4545	.4990	.7061	.7854	.8437	.8866	.9180	.9409	.9576
2	.0361	.1205	.2269	.3388	.4465	.5445	.6306	.7042	.7657	.8163
3	.0033	.0217	.0601	.1169	.1878	.2676	.3513	.4346	.5145	.5886
4	.0002	.0029	.0119	.0306	.0608	.1026	.1550	.2158	.2825	.3526
5	.0000	.0003	.0019	.0063	.0156	.0315	.0553	.0874	.1277	.1755
6	.0000	.0000	.0002	.0011	.0033	.0079	.0162	.0293	.0481	.0732
7	.0000	.0000	.0000	.0001	.0006	.0017	.0040	.0082	.0152	.0258
8	.0000	.0000	.0000	.0000	.0001	.0003	.0008	.0020	.0041	.0078
9	.0000	.0000	.0000	.0000	.0000	.0000	.0001	.0004	.0010	.0020
10	.0000	.0000	.0000	.0000	.0000	.0000	.0000	.0001	.0002	.0005
11	.0000	.0000	.0000	.0000	.0000	.0000	.0000	.0000	.0000	.0001

r \ p	.11	.12	.13	.14	.15	.16	.17	.18	.19	.20
1	.9697	.9784	.9847	.9892	.9924	.9946	.9963	.9974	.9982	.9988
2	.8573	.8900	.9159	.9362	.9520	.9641	.9733	.9803	.9856	.9895
3	.6558	.7153	.7670	.8113	.8486	.8796	.9051	.9259	.9425	.9558
4	.4234	.4929	.5594	.6215	.6783	.7295	.7748	.8144	.8484	.8773
5	.2295	.2882	.3499	.4129	.4755	.5365	.5947	.6491	.6993	.7448
6	.1049	.1431	.1871	.2363	.2894	.3453	.4028	.4605	.5174	.5725
7	.0407	.0606	.0858	.1165	.1526	.1936	.2390	.2880	.3397	.3930
8	.0136	.0221	.0339	.0497	.0698	.0945	.1240	.1582	.1968	.2392
9	.0039	.0069	.0116	.0184	.0278	.0403	.0563	.0763	.1004	.1287
10	.0010	.0019	.0035	.0059	.0097	.0150	.0224	.0323	.0451	.0611
11	.0002	.0005	.0009	.0017	.0029	.0049	.0078	.0120	.0179	.0256
12	.0000	.0001	.0002	.0004	.0008	.0014	.0024	.0040	.0062	.0095
13	.0000	.0000	.0000	.0001	.0002	.0004	.0007	.0011	.0019	.0031
14	.0000	.0000	.0000	.0000	.0000	.0001	.0002	.0003	.0005	.0009
15	.0000	.0000	.0000	.0000	.0000	.0000	.0000	.0001	.0001	.0002
16	.0000	.0000	.0000	.0000	.0000	.0000	.0000	.0000	.0000	.0001

r \ p	.21	.22	.23	.24	.25	.26	.27	.28	.29	.30
1	.9992	.9994	.9996	.9997	.9998	.9999	.9999	.9999	1.0000	1.0000
2	.9924	.9945	.9961	.9972	.9980	.9986	.9990	.9993	.9995	.9997
3	.9663	.9745	.9808	.9857	.9894	.9922	.9943	.9959	.9970	.9979
4	.9016	.9217	.9383	.9517	.9626	.9712	.9780	.9834	.9875	.9907
5	.7854	.8213	.8525	.8793	.9021	.9213	.9373	.9505	.9612	.9698
6	.6249	.6739	.7192	.7604	.7974	.8302	.8590	.8839	.9053	.9234
7	.4470	.5008	.5533	.6039	.6519	.6969	.7384	.7762	.8102	.8405
8	.2850	.3333	.3834	.4345	.4857	.5362	.5853	.6324	.6770	.7186
9	.1611	.1975	.2376	.2807	.3264	.3739	.4226	.4718	.5206	.5685
10	.0806	.1039	.1311	.1620	.1966	.2346	.2756	.3190	.3645	.4112
11	.0357	.0485	.0642	.0833	.1057	.1317	.1613	.1943	.2305	.2696
12	.0140	.0200	.0280	.0381	.0507	.0660	.0845	.1061	.1310	.1593
13	.0049	.0073	.0108	.0155	.0216	.0295	.0395	.0518	.1667	.0845
14	.0015	.0024	.0037	.0056	.0082	.0117	.0164	.0225	.0303	.0401
15	.0004	.0007	.0011	.0018	.0028	.0041	.0061	.0087	.0123	.0169
16	.0001	.0002	.0003	.0005	.0008	.0013	.0020	.0030	.0044	.0064
17	.0000	.0000	.0001	.0001	.0002	.0004	.0006	.0009	.0014	.0021
18	.0000	.0000	.0000	.0000	.0001	.0001	.0001	.0002	.0004	.0006
19	.0000	.0000	.0000	.0000	.0000	.0000	.0000	.0001	.0001	.0002

Table A.1—*Continued*

n = 30 (Continued)

r \ p	.31	.32	.33	.34	.35	.36	.37	.38	.39	.40
1	1.0000	1.0000	1.0000	1.0000	1.0000	1.0000	1.0000	1.0000	1.0000	1.0000
2	.9998	.9999	.9999	.9999	1.0000	1.0000	1.0000	1.0000	1.0000	1.0000
3	.9985	.9989	.9993	.9995	.9997	.9998	.9998	.9999	.9999	1.0000
4	.9931	.9950	.9963	.9974	.9981	.9987	.9991	.9993	.9995	.9997
5	.9768	.9823	.9866	.9899	.9925	.9944	.9959	.9971	.9979	.9985
6	.9386	.9512	.9615	.9700	.9767	.9822	.9864	.9898	.9924	.9943
7	.8671	.8903	.9102	.9271	.9414	.9533	.9631	.9712	.9776	.9828
8	.7570	.7920	.8235	.8515	.8762	.8977	.9163	.9321	.9454	.9565
9	.6148	.6590	.7007	.7395	.7753	.8078	.8371	.8631	.8861	.9060
10	.4586	.5060	.5529	.5985	.6425	.6842	.7235	.7599	.7934	.8237
11	.3112	.3549	.4000	.4460	.4922	.5382	.5833	.6270	.6689	.7085
12	.1909	.2255	.2631	.3031	.3452	.3889	.4337	.4790	.5242	.5689
13	.1053	.1292	.1563	.1865	.2198	.2559	.2945	.3353	.3778	.4215
14	.0520	.0664	.0835	.1034	.1263	.1523	.1813	.2133	.2481	.2855
15	.0229	.0305	.0399	.0514	.0652	.0815	.1006	.1226	.1475	.1754
16	.0090	.0125	.0170	.0228	.0301	.0391	.0501	.0632	.0788	.0971
17	.0031	.0045	.0065	.0090	.0124	.0167	.0222	.0291	.0377	.0481
18	.0010	.0015	.0022	.0032	.0045	.0063	.0088	.0119	.0160	.0212
19	.0003	.0004	.0006	.0010	.0014	.0021	.0031	.0043	.0060	.0083
20	.0001	.0001	.0002	.0003	.0004	.0006	.0009	.0014	.0020	.0029
21	.0000	.0000	.0000	.0001	.0001	.0002	.0002	.0004	.0006	.0009
22	.0000	.0000	.0000	.0000	.0000	.0000	.0001	.0001	.0001	.0002

r \ p	.41	.42	.43	.44	.45	.46	.47	.48	.49	.50
1	1.0000	1.0000	1.0000	1.0000	1.0000	1.0000	1.0000	1.0000	1.0000	1.0000
2	1.0000	1.0000	1.0000	1.0000	1.0000	1.0000	1.0000	1.0000	1.0000	1.0000
3	1.0000	1.0000	1.0000	1.0000	1.0000	1.0000	1.0000	1.0000	1.0000	1.0000
4	.9998	.9999	.9999	.9999	1.0000	1.0000	1.0000	1.0000	1.0000	1.0000
5	.9989	.9993	.9995	.9996	.9998	.9998	.9999	.9999	1.0000	1.0000
6	.9959	.9970	.9978	.9985	.9989	.9992	.9995	.9996	.9998	.9998
7	.9869	.9901	.9926	.9946	.9960	.9971	.9979	.9985	.9990	.9993
8	.9656	.9731	.9792	.9841	.9879	.9909	.9932	.9950	.9964	.9974
9	.9231	.9378	.9501	.9603	.9688	.9757	.9813	.9857	.9892	.9919
10	.8510	.8751	.8964	.9148	.9306	.9440	.9553	.9647	.9724	.9786
11	.7456	.7799	.8112	.8396	.8650	.8874	.9070	.9239	.9384	.9506
12	.6125	.6545	.6945	.7322	.7673	.7996	.8290	.8555	.8790	.8998
13	.4660	.5107	.5551	.5986	.6408	.6813	.7196	.7555	.7888	.8192
14	.3251	.3666	.4095	.4533	.4975	.5417	.5852	.6277	.6687	.7077
15	.2062	.2398	.2760	.3146	.3552	.3973	.4406	.4845	.5285	.5722
16	.1180	.1419	.1687	.1984	.2309	.2661	.3037	.3434	.3849	.4278
17	.0606	.0754	.0928	.1128	.1356	.1613	.1899	.2214	.2556	.2923
18	.0278	.0358	.0456	.0574	.0714	.0878	.1068	.1286	.1533	.1808
19	.0113	.0151	.0199	.0260	.0334	.0426	.0536	.0668	.0822	.1002
20	.0040	.0056	.0077	.0104	.0138	.0183	.0238	.0307	.0391	.0494
21	.0013	.0018	.0026	.0036	.0050	.0069	.0093	.0124	.0164	.0214
22	.0003	.0005	.0008	.0011	.0016	.0022	.0031	.0043	.0060	.0081
23	.0001	.0001	.0002	.0003	.0004	.0006	.0009	.0013	.0019	.0026
24	.0000	.0000	.0000	.0001	.0001	.0001	.0002	.0003	.0005	.0007
25	.0000	.0000	.0000	.0000	.0000	.0000	.0000	.0001	.0001	.0002

n = 40

r \ p	.01	.02	.03	.04	.05	.06	.07	.08	.09	.10
1	.3310	.5543	.7043	.8046	.8715	.9158	.9451	.9644	.9770	.9852
2	.0607	.1905	.3385	.4790	.6009	.7010	.7799	.8406	.8860	.9195
3	.0075	.0457	.1178	.2145	.3233	.4335	.5375	.6306	.7106	.7772
4	.0007	.0082	.0314	.0748	.1382	.2173	.3063	.3993	.4908	.5769
5	.0000	.0012	.0067	.0210	.0480	.0896	.1454	.2132	.2897	.3710
6	.0000	.0001	.0012	.0049	.0139	.0309	.0581	.0967	.1465	.2063
7	.0000	.0000	.0002	.0010	.0034	.0091	.0199	.0376	.0639	.0995
8	.0000	.0000	.0000	.0002	.0007	.0023	.0059	.0127	.0242	.0419
9	.0000	.0000	.0000	.0000	.0001	.0005	.0015	.0037	.0081	.0155
10	.0000	.0000	.0000	.0000	.0000	.0001	.0003	.0010	.0024	.0051
11	.0000	.0000	.0000	.0000	.0000	.0000	.0001	.0002	.0006	.0015
12	.0000	.0000	.0000	.0000	.0000	.0000	.0000	.0000	.0001	.0004
13	.0000	.0000	.0000	.0000	.0000	.0000	.0000	.0000	.0000	.0001

Table A.1—*Continued*

n = 40 (Continued)

r \ p	.11	.12	.13	.14	.15	.16	.17	.18	.19	.20
1	.9905	.9940	.9962	.9976	.9985	.9991	.9994	.9996	.9998	.9999
2	.9438	.9612	.9734	.9820	.9879	.9919	.9947	.9965	.9977	.9985
3	.8312	.8739	.9071	.9324	.9514	.9655	.9757	.9831	.9884	.9921
4	.6548	.7232	.7816	.8302	.8698	.9016	.9265	.9458	.9605	.9715
5	.4532	.5331	.6080	.6762	.7367	.7890	.8333	.8701	.9000	.9241
6	.2738	.3464	.4213	.4958	.5675	.6346	.6958	.7504	.7980	.8387
7	.1445	.1980	.2586	.3245	.3933	.4631	.5316	.5971	.6583	.7141
8	.0668	.0996	.1405	.1890	.2441	.3044	.3682	.4337	.4991	.5629
9	.0272	.0443	.0677	.0980	.1354	.1797	.2301	.2857	.3451	.4069
10	.0098	.0175	.0290	.0453	.0672	.0952	.1296	.1702	.2167	.2682
11	.0032	.0062	.0111	.0188	.0299	.0454	.0657	.0916	.1233	.1608
12	.0009	.0019	.0038	.0070	.0120	.0194	.0301	.0446	.0636	.0875
13	.0002	.0005	.0012	.0023	.0043	.0075	.0124	.0196	.0297	.0432
14	.0001	.0001	.0003	.0007	.0014	.0026	.0047	.0078	.0126	.0194
15	.0000	.0000	.0001	.0002	.0004	.0008	.0016	.0028	.0048	.0079
16	.0000	.0000	.0000	.0000	.0001	.0002	.0005	.0009	.0017	.0029
17	.0000	.0000	.0000	.0000	.0000	.0001	.0001	.0003	.0005	.0010
18	.0000	.0000	.0000	.0000	.0000	.0000	.0000	.0001	.0002	.0003
19	.0000	.0000	.0000	.0000	.0000	.0000	.0000	.0000	.0000	.0001

r \ p	.21	.22	.23	.24	.25	.26	.27	.28	.29	.30
1	.9999	1.0000	1.0000	1.0000	1.0000	1.0000	1.0000	1.0000	1.0000	1.0000
2	.9991	.9994	.9996	.9998	.9999	.9999	.9999	1.0000	1.0000	1.0000
3	.9946	.9964	.9976	.9984	.9990	.9993	.9996	.9997	.9998	.9999
4	.9797	.9857	.9900	.9931	.9953	.9968	.9979	.9986	.9991	.9994
5	.9430	.9578	.9691	.9776	.9840	.9886	.9920	.9945	.9962	.9974
6	.8729	.9011	.9240	.9423	.9567	.9679	.9765	.9830	.9878	.9914
7	.7640	.8078	.8454	.8773	.9038	.9255	.9430	.9569	.9678	.9762
8	.6235	.6799	.7314	.7775	.8180	.8530	.8828	.9076	.9281	.9447
9	.4694	.5312	.5910	.6476	.7002	.7480	.7909	.8286	.8612	.8890
10	.3238	.3821	.4419	.5017	.5605	.6169	.6701	.7193	.7641	.8041
11	.2038	.2517	.3038	.3589	.4161	.4740	.5315	.5875	.6410	.6913
12	.1167	.1514	.1912	.2359	.2849	.3371	.3918	.4478	.5040	.5594
13	.0609	.0830	.1100	.1421	.1791	.2209	.2669	.3165	.3687	.4228
14	.0289	.0415	.0578	.0782	.1032	.1329	.1674	.2065	.2498	.2968
15	.0124	.0189	.0277	.0394	.0544	.0733	.0964	.1240	.1560	.1926
16	.0049	.0078	.0121	.0181	.0262	.0370	.0509	.0683	.0897	.1151
17	.0017	.0030	.0048	.0076	.0116	.0171	.0246	.0345	.0473	.0633
18	.0006	.0010	.0018	.0029	.0047	.0072	.0109	.0160	.0229	.0320
19	.0002	.0003	.0006	.0010	.0017	.0028	.0044	.0068	.0101	.0148
20	.0000	.0001	.0002	.0003	.0006	.0010	.0016	.0026	.0041	.0063
21	.0000	.0000	.0000	.0001	.0002	.0003	.0005	.0009	.0015	.0024
22	.0000	.0000	.0000	.0000	.0000	.0001	.0002	.0003	.0005	.0009
23	.0000	.0000	.0000	.0000	.0000	.0000	.0000	.0001	.0002	.0003
24	.0000	.0000	.0000	.0000	.0000	.0000	.0000	.0000	.0000	.0001

r \ p	.31	.32	.33	.34	.35	.36	.37	.38	.39	.40
1	1.0000	1.0000	1.0000	1.0000	1.0000	1.0000	1.0000	1.0000	1.0000	1.0000
2	1.0000	1.0000	1.0000	1.0000	1.0000	1.0000	1.0000	1.0000	1.0000	1.0000
3	.9999	1.0000	1.0000	1.0000	1.0000	1.0000	1.0000	1.0000	1.0000	1.0000
4	.9996	.9998	.9998	.9999	.9999	1.0000	1.0000	1.0000	1.0000	1.0000
5	.9983	.9989	.9993	.9995	.9997	.9998	.9999	.9999	1.0000	1.0000
6	.9940	.9958	.9971	.9981	.9987	.9991	.9994	.9996	.9998	.9999
7	.9827	.9875	.9911	.9937	.9956	.9970	.9980	.9986	.9991	.9994
8	.9580	.9685	.9766	.9829	.9876	.9911	.9937	.9956	.9970	.9979
9	.9123	.9315	.9472	.9598	.9697	.9775	.9835	.9880	.9914	.9939
10	.8393	.8697	.8957	.9175	.9356	.9503	.9621	.9715	.9788	.9844
11	.7376	.7796	.8171	.8500	.8785	.9028	.9232	.9400	.9537	.9648
12	.6130	.6639	.7115	.7552	.7947	.8299	.8608	.8874	.9101	.9291
13	.4777	.5323	.5857	.6371	.6857	.7309	.7722	.8095	.8426	.8715
14	.3467	.3989	.4524	.5061	.5592	.6109	.6602	.7067	.7497	.7888
15	.2333	.2779	.3257	.3759	.4279	.4807	.5334	.5851	.6351	.6826
16	.1450	.1791	.2175	.2597	.3054	.3538	.4043	.4560	.5081	.5598
17	.0830	.1065	.1343	.1662	.2022	.2422	.2858	.3323	.3813	.4319
18	.0436	.0583	.0764	.0981	.1239	.1536	.1875	.2253	.2668	.3115
19	.0210	.0293	.0399	.0534	.0699	.0900	.1138	.1415	.1732	.2089
20	.0093	.0135	.0192	.0266	.0363	.0485	.0636	.0820	.1040	.1298

Table A.1—*Continued*

n = 40 (Continued)

r \ p	.31	.32	.33	.34	.35	.36	.37	.38	.39	.40
21	.0038	.0057	.0084	.0122	.0173	.0240	.0327	.0438	.0575	.0744
22	.0014	.0022	.0034	.0051	.0075	.0109	.0154	.0214	.0292	.0392
23	.0005	.0008	.0012	.0020	.0030	.0045	.0066	.0096	.0136	.0189
24	.0001	.0002	.0004	.0007	.0011	.0017	.0026	.0039	.0058	.0083
25	.0000	.0001	.0001	.0002	.0004	.0006	.0009	.0015	.0022	.0034
26	.0000	.0000	.0000	.0001	.0001	.0002	.0003	.0005	.0008	.0012
27	.0000	.0000	.0000	.0000	.0000	.0001	.0001	.0002	.0002	.0004
28	.0000	.0000	.0000	.0000	.0000	.0000	.0000	.0000	.0001	.0001

r \ p	.41	.42	.43	.44	.45	.46	.47	.48	.49	.50
1	1.0000	1.0000	1.0000	1.0000	1.0000	1.0000	1.0000	1.0000	1.0000	1.0000
2	1.0000	1.0000	1.0000	1.0000	1.0000	1.0000	1.0000	1.0000	1.0000	1.0000
3	1.0000	1.0000	1.0000	1.0000	1.0000	1.0000	1.0000	1.0000	1.0000	1.0000
4	1.0000	1.0000	1.0000	1.0000	1.0000	1.0000	1.0000	1.0000	1.0000	1.0000
5	1.0000	1.0000	1.0000	1.0000	1.0000	1.0000	1.0000	1.0000	1.0000	1.0000
6	.9999	.9999	1.0000	1.0000	1.0000	1.0000	1.0000	1.0000	1.0000	1.0000
7	.9996	.9998	.9998	.9999	.9999	1.0000	1.0000	1.0000	1.0000	1.0000
8	.9986	.9991	.9994	.9996	.9998	.9998	.9999	.9999	1.0000	1.0000
9	.9958	.9971	.9980	.9987	.9991	.9994	.9996	.9998	.9999	.9999
10	.9887	.9919	.9943	.9960	.9973	.9981	.9988	.9992	.9995	.9997
11	.9735	.9803	.9856	.9896	.9926	.9948	.9964	.9975	.9983	.9989
12	.9447	.9575	.9677	.9758	.9821	.9869	.9906	.9933	.9953	.9968
13	.8964	.9175	.9351	.9496	.9614	.9708	.9782	.9840	.9884	.9917
14	.8240	.8551	.8821	.9053	.9249	.9413	.9546	.9654	.9740	.9808
15	.7270	.7679	.8051	.8382	.8674	.8927	.9143	.9324	.9474	.9597
16	.6102	.6586	.7043	.7468	.7858	.8209	.8522	.8795	.9031	.9231
17	.4833	.5348	.5855	.6346	.6815	.7255	.7662	.8033	.8365	.8659
18	.3589	.4083	.4590	.5101	.5609	.6107	.6585	.7039	.7463	.7852
19	.2484	.2912	.3370	.3851	.4349	.4857	.5365	.5867	.6354	.6821
20	.1594	.1930	.2305	.2714	.3156	.3624	.4112	.4614	.5122	.5627
21	.0946	.1184	.1461	.1776	.2130	.2521	.2946	.3400	.3878	.4373
22	.0516	.0669	.0855	.1074	.1331	.1627	.1961	.2333	.2740	.3179
23	.0259	.0348	.0460	.0598	.0767	.0969	.1206	.1482	.1796	.2148
24	.0118	.0165	.0226	.0305	.0405	.0530	.0683	.0867	.1086	.1341
25	.0049	.0072	.0102	.0142	.0196	.0265	.0354	.0465	.0602	.0769
26	.0019	.0028	.0042	.0060	.0086	.0121	.0167	.0228	.0305	.0403
27	.0006	.0010	.0015	.0023	.0034	.0050	.0072	.0101	.0140	.0192
28	.0002	.0003	.0005	.0008	.0012	.0019	.0028	.0041	.0058	.0083
29	.0001	.0001	.0002	.0002	.0004	.0006	.0010	.0015	.0022	.0032
30	.0000	.0000	.0000	.0001	.0001	.0002	.0003	.0005	.0007	.0011
31	.0000	.0000	.0000	.0000	.0000	.0000	.0001	.0001	.0002	.0003
32	.0000	.0000	.0000	.0000	.0000	.0000	.0000	.0000	.0001	.0001

n = 50

r \ p	.01	.02	.03	.04	.05	.06	.07	.08	.09	.10
1	.3950	.6358	.7819	.8701	.9231	.9547	.9734	.9845	.9910	.9948
2	.0894	.2642	.4447	.5995	.7206	.8100	.8735	.9173	.9468	.9662
3	.0138	.0784	.1892	.3233	.4595	.5838	.6892	.7740	.8395	.8883
4	.0016	.0178	.0628	.1391	.2396	.3527	.4673	.5747	.6697	.7497
5	.0001	.0032	.0168	.0490	.1036	.1794	.2710	.3711	.4723	.5688
6	.0000	.0005	.0037	.0144	.0378	.0776	.1350	.2081	.2928	.3839
7	.0000	.0001	.0007	.0036	.0118	.0289	.0583	.1019	.1596	.2298
8	.0000	.0000	.0001	.0008	.0032	.0094	.0220	.0438	.0768	.1221
9	.0000	.0000	.0000	.0001	.0008	.0027	.0073	.0167	.0328	.0579
10	.0000	.0000	.0000	.0000	.0002	.0007	.0022	.0056	.0125	.0245
11	.0000	.0000	.0000	.0000	.0000	.0002	.0006	.0017	.0043	.0094
12	.0000	.0000	.0000	.0000	.0000	.0000	.0001	.0005	.0013	.0032
13	.0000	.0000	.0000	.0000	.0000	.0000	.0000	.0001	.0004	.0010
14	.0000	.0000	.0000	.0000	.0000	.0000	.0000	.0000	.0001	.0003
15	.0000	.0000	.0000	.0000	.0000	.0000	.0000	.0000	.0000	.0001

r \ p	.11	.12	.13	.14	.15	.16	.17	.18	.19	.20
1	.9971	.9983	.9991	.9995	.9997	.9998	.9999	1.0000	1.0000	1.0000
2	.9788	.9869	.9920	.9951	.9971	.9983	.9990	.9994	.9997	.9998
3	.9237	.9487	.9661	.9779	.9858	.9910	.9944	.9965	.9979	.9987
4	.8146	.8655	.9042	.9330	.9540	.9688	.9792	.9863	.9912	.9943
5	.6562	.7320	.7956	.8472	.8879	.9192	.9428	.9601	.9726	.9815

Table A.1—*Continued*

n = 50 (Continued)

r \ p	.11	.12	.13	.14	.15	.16	.17	.18	.19	.20
6	.4760	.5647	.6463	.7186	.7806	.8323	.8741	.9071	.9327	.9520
7	.3091	.3935	.4789	.5616	.6387	.7081	.7686	.8199	.8624	.8966
8	.1793	.2467	.3217	.4010	.4812	.5594	.6328	.6996	.7587	.8096
9	.0932	.1392	.1955	.2605	.3319	.4071	.4832	.5576	.6280	.6927
10	.0435	.0708	.1074	.1537	.2089	.2718	.3403	.4122	.4849	.5563
11	.0183	.0325	.0535	.0824	.1199	.1661	.2203	.2813	.3473	.4164
12	.0069	.0135	.0242	.0402	.0628	.0929	.1309	.1768	.2300	.2893
13	.0024	.0051	.0100	.0179	.0301	.0475	.0714	.1022	.1405	.1861
14	.0008	.0018	.0037	.0073	.0132	.0223	.0357	.0544	.0791	.1106
15	.0002	.0006	.0013	.0027	.0053	.0096	.0164	.0266	.0411	.0607
16	.0001	.0002	.0004	.0009	.0020	.0038	.0070	.0120	.0197	.0308
17	.0000	.0000	.0001	.0003	.0007	.0014	.0027	.0050	.0087	.0144
18	.0000	.0000	.0000	.0001	.0002	.0005	.0010	.0019	.0036	.0063
19	.0000	.0000	.0000	.0000	.0001	.0001	.0003	.0007	.0013	.0025
20	.0000	.0000	.0000	.0000	.0000	.0000	.0001	.0002	.0005	.0009
21	.0000	.0000	.0000	.0000	.0000	.0000	.0000	.0001	.0002	.0003
22	.0000	.0000	.0000	.0000	.0000	.0000	.0000	.0000	.0000	.0001

r \ p	.21	.22	.23	.24	.25	.26	.27	.28	.29	.30
1	1.0000	1.0000	1.0000	1.0000	1.0000	1.0000	1.0000	1.0000	1.0000	1.0000
2	.9999	.9999	1.0000	1.0000	1.0000	1.0000	1.0000	1.0000	1.0000	1.0000
3	.9992	.9995	.9997	.9998	.9999	1.0000	1.0000	1.0000	1.0000	1.0000
4	.9964	.9978	.9986	.9992	.9995	.9997	.9998	.9999	.9999	1.0000
5	.9877	.9919	.9948	.9967	.9979	.9987	.9992	.9995	.9997	.9998
6	.9963	.9767	.9841	.9893	.9930	.9954	.9970	.9981	.9988	.9993
7	.9236	.9445	.9603	.9720	.9806	.9868	.9911	.9941	.9961	.9975
8	.8523	.8874	.9156	.9377	.9547	.9676	.9772	.9842	.9892	.9927
9	.7505	.8009	.8437	.8794	.9084	.9316	.9497	.9635	.9740	.9817
10	.6241	.6870	.7436	.7934	.8363	.8724	.9021	.9260	.9450	.9598
11	.4864	.5552	.6210	.6822	.7378	.7871	.8299	.8663	.8965	.9211
12	.3533	.4201	.4878	.5544	.6184	.6782	.7329	.7817	.8244	.8610
13	.2383	.2963	.3585	.4233	.4890	.5539	.6163	.6749	.7287	.7771
14	.1490	.1942	.2456	.3023	.3630	.4261	.4901	.5534	.6145	.6721
15	.0862	.1181	.1565	.2013	.2519	.3075	.3669	.4286	.4912	.5532
16	.0462	.0665	.0926	.1247	.1631	.2075	.2575	.3121	.3703	.4308
17	.0229	.0347	.0508	.0718	.0983	.1306	.1689	.2130	.2623	.3161
18	.0105	.0168	.0259	.0384	.0551	.0766	.1034	.1359	.1741	.2178
19	.0045	.0075	.0122	.0191	.0287	.0418	.0590	.0809	.1080	.1406
20	.0018	.0031	.0054	.0088	.0139	.0212	.0314	.0449	.0626	.0848
21	.0006	.0012	.0022	.0038	.0063	.0100	.0155	.0232	.0338	.0478
22	.0002	.0004	.0008	.0015	.0026	.0044	.0071	.0112	.0170	.0251
23	.0001	.0001	.0003	.0006	.0010	.0018	.0031	.0050	.0080	.0123
24	.0000	.0000	.0001	.0002	.0004	.0007	.0012	.0021	.0035	.0056
25	.0000	.0000	.0000	.0001	.0001	.0002	.0004	.0008	.0014	.0024
26	.0000	.0000	.0000	.0000	.0000	.0001	.0002	.0003	.0005	.0009
27	.0000	.0000	.0000	.0000	.0000	.0000	.0001	.0001	.0002	.0003
28	.0000	.0000	.0000	.0000	.0000	.0000	.0000	.0000	.0001	.0001

r \ p	.31	.32	.33	.34	.35	.36	.37	.38	.39	.40
1	1.0000	1.0000	1.0000	1.0000	1.0000	1.0000	1.0000	1.0000	1.0000	1.0000
2	1.0000	1.0000	1.0000	1.0000	1.0000	1.0000	1.0000	1.0000	1.0000	1.0000
3	1.0000	1.0000	1.0000	1.0000	1.0000	1.0000	1.0000	1.0000	1.0000	1.0000
4	1.0000	1.0000	1.0000	1.0000	1.0000	1.0000	1.0000	1.0000	1.0000	1.0000
5	.9999	.9999	1.0000	1.0000	1.0000	1.0000	1.0000	1.0000	1.0000	1.0000
6	.9996	.9997	.9998	.9999	.9999	1.0000	1.0000	1.0000	1.0000	1.0000
7	.9984	.9990	.9994	.9996	.9998	.9999	.9999	1.0000	1.0000	1.0000
8	.9952	.9969	.9980	.9987	.9992	.9995	.9997	.9998	.9999	.9999
9	.9874	.9914	.9942	.9962	.9975	.9984	.9990	.9994	.9996	.9998
10	.9710	.9794	.9856	.9901	.9933	.9955	.9971	.9981	.9988	.9992
11	.9409	.9563	.9683	.9773	.9840	.9889	.9924	.9949	.9966	.9978
12	.8916	.9168	.9371	.9533	.9658	.9753	.9825	.9878	.9916	.9943
13	.8197	.8564	.8874	.9130	.9339	.9505	.9635	.9736	.9811	.9867
14	.7253	.7732	.8157	.8524	.8837	.9097	.9310	.9481	.9616	.9720
15	.6131	.6698	.7223	.7699	.8122	.8491	.8805	.9069	.9286	.9460
16	.4922	.5530	.6120	.6679	.7199	.7672	.8094	.8462	.8779	.9045
17	.3734	.4328	.4931	.5530	.6111	.6664	.7179	.7649	.8070	.8439
18	.2666	.3197	.3760	.4346	.4940	.5531	.6105	.6653	.7164	.7631
19	.1786	.2220	.2703	.3227	.3784	.4362	.4949	.5533	.6101	.6644
20	.1121	.1447	.1826	.2257	.2736	.3255	.3805	.4376	.4957	.5535

Table A.1—*Continued*

n = 50 (Continued)

r \ p	.31	.32	.33	.34	.35	.36	.37	.38	.39	.40
21	.0657	.0882	.1156	.1482	.1861	.2289	.2764	.3278	.3824	.4390
22	.0360	.0503	.0685	.0912	.1187	.1513	.1890	.2317	.2788	.3299
23	.0184	.0267	.0379	.0525	.0710	.0938	.1214	.1540	.1916	.2340
24	.0087	.0133	.0196	.0282	.0396	.0544	.0730	.0960	.1236	.1562
25	.0039	.0061	.0094	.0141	.0207	.0295	.0411	.0560	.0748	.0978
26	.0016	.0026	.0042	.0066	.0100	.0149	.0216	.0306	.0423	.0573
27	.0006	.0011	.0018	.0029	.0045	.0070	.0106	.0155	.0223	.0314
28	.0002	.0004	.0007	.0012	.0019	.0031	.0048	.0074	.0110	.0160
29	.0001	.0001	.0002	.0004	.0007	.0012	.0020	.0032	.0050	.0076
30	.0000	.0000	.0001	.0002	.0003	.0005	.0008	.0013	.0021	.0034
31	.0000	.0000	.0000	.0000	.0001	.0002	.0003	.0005	.0008	.0014
32	.0000	.0000	.0000	.0000	.0000	.0001	.0001	.0002	.0003	.0005
33	.0000	.0000	.0000	.0000	.0000	.0000	.0000	.0001	.0001	.0002
34	.0000	.0000	.0000	.0000	.0000	.0000	.0000	.0000	.0000	.0001

r \ p	.41	.42	.43	.44	.45	.46	.47	.48	.49	.50
1	1.0000	1.0000	1.0000	1.0000	1.0000	1.0000	1.0000	1.0000	1.0000	1.0000
2	1.0000	1.0000	1.0000	1.0000	1.0000	1.0000	1.0000	1.0000	1.0000	1.0000
3	1.0000	1.0000	1.0000	1.0000	1.0000	1.0000	1.0000	1.0000	1.0000	1.0000
4	1.0000	1.0000	1.0000	1.0000	1.0000	1.0000	1.0000	1.0000	1.0000	1.0000
5	1.0000	1.0000	1.0000	1.0000	1.0000	1.0000	1.0000	1.0000	1.0000	1.0000
6	1.0000	1.0000	1.0000	1.0000	1.0000	1.0000	1.0000	1.0000	1.0000	1.0000
7	1.0000	1.0000	1.0000	1.0000	1.0000	1.0000	1.0000	1.0000	1.0000	1.0000
8	1.0000	1.0000	1.0000	1.0000	1.0000	1.0000	1.0000	1.0000	1.0000	1.0000
9	.9999	.9999	1.0000	1.0000	1.0000	1.0000	1.0000	1.0000	1.0000	1.0000
10	.9995	.9997	.9998	.9999	.9999	1.0000	1.0000	1.0000	1.0000	1.0000
11	.9986	.9991	.9994	.9997	.9998	.9999	.9999	1.0000	1.0000	1.0000
12	.9962	.9975	.9984	.9990	.9994	.9996	.9998	.9999	.9999	1.0000
13	.9908	.9938	.9958	.9973	.9982	.9989	.9993	.9996	.9997	.9998
14	.9799	.9858	.9902	.9933	.9955	.9970	.9981	.9988	.9992	.9995
15	.9599	.9707	.9789	.9851	.9896	.9929	.9952	.9968	.9980	.9987
16	.9265	.9443	.9585	.9696	.9780	.9844	.9892	.9926	.9950	.9967
17	.8757	.9025	.9248	.9429	.9573	.9687	.9774	.9839	.9888	.9923
18	.8051	.8421	.8740	.9010	.9235	.9418	.9565	.9680	.9769	.9836
19	.7151	.7617	.8037	.8406	.8727	.8998	.9225	.9410	.9559	.9675
20	.6099	.6638	.7143	.7608	.8026	.8396	.8718	.8991	.9219	.9405
21	.4965	.5539	.6099	.6635	.7138	.7602	.8020	.8391	.8713	.8987
22	.3840	.4402	.4973	.5543	.6100	.6634	.7137	.7599	.8018	.8389
23	.2807	.3316	.3854	.4412	.4981	.5548	.6104	.6636	.7138	.7601
24	.1936	.2359	.2826	.3331	.3866	.4422	.4989	.5554	.6109	.6641
25	.1255	.1580	.1953	.2375	.2840	.3343	.3876	.4431	.4996	.5561
26	.0762	.0992	.1269	.1593	.1966	.2386	.2850	.3352	.3885	.4439
27	.0432	.0584	.0772	.1003	.1279	.1603	.1975	.2395	.2858	.3359
28	.0229	.0321	.0439	.0591	.0780	.1010	.1286	.1609	.1981	.2399
29	.0113	.0164	.0233	.0325	.0444	.0595	.0784	.1013	.1289	.1611
30	.0052	.0078	.0115	.0166	.0235	.0327	.0446	.0596	.0784	.1013
31	.0022	.0034	.0053	.0079	.0116	.0167	.0236	.0327	.0445	.0595
32	.0009	.0014	.0022	.0035	.0053	.0079	.0116	.0166	.0234	.0325
33	.0003	.0005	.0009	.0014	.0022	.0035	.0053	.0078	.0114	.0164
34	.0001	.0002	.0003	.0005	.0009	.0014	.0022	.0034	.0052	.0077
35	.0000	.0001	.0001	.0002	.0003	.0005	.0009	.0014	.0021	.0033
36	.0000	.0000	.0000	.0001	.0001	.0002	.0003	.0005	.0008	.0013
37	.0000	.0000	.0000	.0000	.0000	.0001	.0001	.0002	.0003	.0005
38	.0000	.0000	.0000	.0000	.0000	.0000	.0000	.0001	.0001	.0002

n = 75

r \ p	.01	.02	.03	.04	.05	.06	.07	.08	.09	.10
1	.5294	.7802	.8982	.9532	.9787	.9903	.9957	.9981	.9992	.9996
2	.1729	.4439	.6620	.8069	.8944	.9441	.9712	.9855	.9929	.9965
3	.0397	.1899	.3917	.5814	.7303	.8350	.9032	.9452	.9699	.9839
4	.0069	.0637	.1882	.3527	.5202	.6655	.7786	.8598	.9145	.9496
5	.0010	.0174	.0750	.1812	.3211	.4708	.6098	.7261	.8160	.8811
6	.0001	.0040	.0253	.0798	.1724	.2943	.4294	.5610	.6776	.7729
7	.0000	.0008	.0073	.0305	.0810	.1629	.2709	.3936	.5179	.6327
8	.0000	.0001	.0019	.0102	.0336	.0802	.1534	.2501	.3622	.4792
9	.0000	.0000	.0004	.0030	.0124	.0353	.0781	.1440	.2313	.3342
10	.0000	.0000	.0001	.0008	.0041	.0140	.0360	.0753	.1350	.2142

Table A.1—*Continued*

n = 75 (Continued)

r \ p	.01	.02	.03	.04	.05	.06	.07	.08	.09	.10
11	.0000	.0000	.0000	.0002	.0012	.0050	.0151	.0359	.0721	.1263
12	.0000	.0000	.0000	.0000	.0003	.0016	.0057	.0157	.0353	.0685
13	.0000	.0000	.0000	.0000	.0001	.0005	.0020	.0063	.0159	.0343
14	.0000	.0000	.0000	.0000	.0000	.0001	.0006	.0023	.0066	.0159
15	.0000	.0000	.0000	.0000	.0000	.0000	.0002	.0008	.0025	.0068
16	.0000	.0000	.0000	.0000	.0000	.0000	.0001	.0002	.0009	.0027
17	.0000	.0000	.0000	.0000	.0000	.0000	.0000	.0001	.0003	.0010
18	.0000	.0000	.0000	.0000	.0000	.0000	.0000	.0000	.0001	.0003
19	.0000	.0000	.0000	.0000	.0000	.0000	.0000	.0000	.0000	.0001

r \ p	.11	.12	.13	.14	.15	.16	.17	.18	.19	.20
1	.9998	.9999	1.0000	1.0000	1.0000	1.0000	1.0000	1.0000	1.0000	1.0000
2	.9984	.9992	.9996	.9998	.9999	1.0000	1.0000	1.0000	1.0000	1.0000
3	.9916	.9957	.9978	.9989	.9995	.9998	.9999	.9999	1.0000	1.0000
4	.9712	.9839	.9913	.9954	.9976	.9988	.9994	.9997	.9999	.9999
5	.9258	.9551	.9736	.9849	.9916	.9954	.9976	.9987	.9994	.9997
6	.8461	.8993	.9362	.9608	.9766	.9864	.9923	.9957	.9977	.9988
7	.7312	.8105	.8710	.9150	.9456	.9662	.9796	.9880	.9931	.9961
8	.5913	.6912	.7749	.8414	.8918	.9284	.9540	.9712	.9825	.9896
9	.4442	.5528	.6529	.7397	.8111	.8672	.9094	.9400	.9613	.9757
10	.3090	.4124	.5171	.6164	.7051	.7804	.8415	.8889	.9243	.9499
11	.1986	.2860	.3833	.4839	.5816	.6713	.7496	.8149	.8671	.9072
12	.1180	.1842	.2651	.3564	.4528	.5485	.6385	.7190	.7878	.8443
13	.0649	.1101	.1708	.2458	.3316	.4237	.5170	.6066	.6886	.7603
14	.0330	.0612	.1026	.1585	.2279	.3086	.3965	.4871	.5758	.6586
15	.0156	.0316	.0575	.0955	.1469	.2114	.2872	.3709	.4586	.5460
16	.0069	.0152	.0301	.0539	.0888	.1362	.1961	.2672	.3468	.4315
17	.0028	.0068	.0147	.0284	.0503	.0824	.1261	.1818	.2485	.3241
18	.0011	.0029	.0067	.0141	.0268	.0469	.0764	.1168	.1685	.2310
19	.0004	.0011	.0029	.0065	.0134	.0251	.0436	.0707	.1080	.1560
20	.0001	.0004	.0012	.0028	.0063	.0126	.0235	.0404	.0654	.0997
21	.0000	.0001	.0004	.0012	.0028	.0060	.0119	.0218	.0374	.0603
22	.0000	.0000	.0002	.0004	.0012	.0027	.0057	.0111	.0202	.0345
23	.0000	.0000	.0001	.0002	.0005	.0011	.0026	.0054	.0104	.0187
24	.0000	.0000	.0000	.0001	.0002	.0005	.0011	.0024	.0050	.0096
25	.0000	.0000	.0000	.0000	.0001	.0002	.0004	.0010	.0023	.0047
26	.0000	.0000	.0000	.0000	.0000	.0001	.0002	.0004	.0010	.0021
27	.0000	.0000	.0000	.0000	.0000	.0000	.0001	.0002	.0004	.0009
28	.0000	.0000	.0000	.0000	.0000	.0000	.0000	.0001	.0002	.0004
29	.0000	.0000	.0000	.0000	.0000	.0000	.0000	.0000	.0001	.0002
30	.0000	.0000	.0000	.0000	.0000	.0000	.0000	.0000	.0000	.0001

r \ p	.21	.22	.23	.24	.25	.26	.27	.28	.29	.30
1	1.0000	1.0000	1.0000	1.0000	1.0000	1.0000	1.0000	1.0000	1.0000	1.0000
2	1.0000	1.0000	1.0000	1.0000	1.0000	1.0000	1.0000	1.0000	1.0000	1.0000
3	1.0000	1.0000	1.0000	1.0000	1.0000	1.0000	1.0000	1.0000	1.0000	1.0000
4	1.0000	1.0000	1.0000	1.0000	1.0000	1.0000	1.0000	1.0000	1.0000	1.0000
5	.9998	.9999	1.0000	1.0000	1.0000	1.0000	1.0000	1.0000	1.0000	1.0000
6	.9994	.9997	.9998	.9999	1.0000	1.0000	1.0000	1.0000	1.0000	1.0000
7	.9979	.9989	.9994	.9997	.9998	.9999	1.0000	1.0000	1.0000	1.0000
8	.9940	.9966	.9981	.9990	.9995	.9997	.9999	.9999	1.0000	1.0000
9	.9851	.9911	.9948	.9971	.9984	.9991	.9995	.9998	.9999	.9999
10	.9677	.9797	.9875	.9925	.9956	.9975	.9986	.9992	.9996	.9998
11	.9370	.9583	.9732	.9831	.9897	.9938	.9964	.9979	.9988	.9994
12	.8889	.9228	.9478	.9656	.9779	.9861	.9915	.9949	.9970	.9983
13	.8206	.8693	.9074	.9360	.9569	.9717	.9819	.9887	.9931	.9959
14	.7327	.7963	.8488	.8908	.9231	.9472	.9647	.9769	.9853	.9909
15	.6291	.7050	.7714	.8275	.8731	.9091	.9365	.9567	.9712	.9813
16	.5172	.6003	.6774	.7462	.8054	.8546	.8940	.9247	.9478	.9647
17	.4057	.4895	.5720	.6500	.7208	.7827	.8352	.8780	.9120	.9380
18	.3028	.3811	.4628	.5445	.6228	.6952	.7596	.8150	.8611	.8983
19	.2146	.2826	.3577	.4372	.5176	.5960	.6696	.7361	.7942	.8434
20	.1443	.1992	.2636	.3355	.4125	.4915	.5696	.6440	.7122	.7729
21	.0920	.1334	.1848	.2456	.3143	.3887	.4662	.5437	.6185	.6882
22	.0556	.0848	.1232	.1712	.2286	.2941	.3659	.4415	.5182	.5932
23	.0318	.0511	.0780	.1136	.1585	.2125	.2749	.3440	.4177	.4933
24	.0173	.0292	.0469	.0717	.1046	.1465	.1974	.2567	.3230	.3945
25	.0089	.0158	.0268	.0430	.0657	.0962	.1352	.1830	.2393	.3029

Table A.1—*Continued*

n = 75 (Continued)

r \ p	.21	.22	.23	.24	.25	.26	.27	.28	.29	.30
26	.0043	.0082	.0145	.0245	.0393	.0602	.0883	.1246	.1695	.2227
27	.0020	.0040	.0075	.0132	.0223	.0358	.0550	.0809	.1147	.1567
28	.0009	.0018	.0036	.0068	.0120	.0203	.0326	.0501	.0740	.1053
29	.0004	.0008	.0017	.0033	.0062	.0109	.0184	.0296	.0456	.0676
30	.0001	.0003	.0007	.0015	.0030	.0056	.0099	.0166	.0268	.0414
31	.0001	.0001	.0003	.0007	.0014	.0027	.0050	.0089	.0150	.0242
32	.0000	.0001	.0001	.0003	.0006	.0013	.0025	.0045	.0080	.0134
33	.0000	.0000	.0000	.0001	.0003	.0006	.0011	.0022	.0040	.0071
34	.0000	.0000	.0000	.0000	.0001	.0002	.0005	.0010	.0020	.0036
35	.0000	.0000	.0000	.0000	.0000	.0001	.0002	.0004	.0009	.0017
36	.0000	.0000	.0000	.0000	.0000	.0000	.0001	.0002	.0004	.0008
37	.0000	.0000	.0000	.0000	.0000	.0000	.0000	.0001	.0002	.0003
38	.0000	.0000	.0000	.0000	.0000	.0000	.0000	.0000	.0001	.0001
39	.0000	.0000	.0000	.0000	.0000	.0000	.0000	.0000	.0000	.0001

r \ p	.31	.32	.33	.34	.35	.36	.37	.38	.39	.40
1	1.0000	1.0000	1.0000	1.0000	1.0000	1.0000	1.0000	1.0000	1.0000	1.0000
2	1.0000	1.0000	1.0000	1.0000	1.0000	1.0000	1.0000	1.0000	1.0000	1.0000
3	1.0000	1.0000	1.0000	1.0000	1.0000	1.0000	1.0000	1.0000	1.0000	1.0000
4	1.0000	1.0000	1.0000	1.0000	1.0000	1.0000	1.0000	1.0000	1.0000	1.0000
5	1.0000	1.0000	1.0000	1.0000	1.0000	1.0000	1.0000	1.0000	1.0000	1.0000
6	1.0000	1.0000	1.0000	1.0000	1.0000	1.0000	1.0000	1.0000	1.0000	1.0000
7	1.0000	1.0000	1.0000	1.0000	1.0000	1.0000	1.0000	1.0000	1.0000	1.0000
8	1.0000	1.0000	1.0000	1.0000	1.0000	1.0000	1.0000	1.0000	1.0000	1.0000
9	1.0000	1.0000	1.0000	1.0000	1.0000	1.0000	1.0000	1.0000	1.0000	1.0000
10	.9999	.9999	1.0000	1.0000	1.0000	1.0000	1.0000	1.0000	1.0000	1.0000
11	.9997	.9998	.9999	1.0000	1.0000	1.0000	1.0000	1.0000	1.0000	1.0000
12	.9991	.9995	.9997	.9999	.9999	1.0000	1.0000	1.0000	1.0000	1.0000
13	.9976	.9986	.9992	.9996	.9998	.9999	.9999	1.0000	1.0000	1.0000
14	.9945	.9967	.9981	.9989	.9994	.9997	.9998	.9999	1.0000	1.0000
15	.9882	.9927	.9956	.9974	.9985	.9992	.9995	.9998	.9999	.9999
16	.9767	.9850	.9906	.9942	.9965	.9980	.9988	.9994	.9997	.9998
17	.9574	.9714	.9813	.9881	.9926	.9955	.9973	.9984	.9991	.9995
18	.9273	.9493	.9655	.9771	.9851	.9906	.9942	.9965	.9979	.9988
19	.8837	.9157	.9404	.9588	.9723	.9818	.9883	.9927	.9955	.9973
20	.8249	.8683	.9033	.9306	.9515	.9669	.9779	.9856	.9909	.9944
21	.7510	.8058	.8521	.8900	.9201	.9434	.9608	.9736	.9826	.9888
22	.6640	.7288	.7860	.8351	.8759	.9088	.9346	.9542	.9687	.9791
23	.5681	.6398	.7062	.7657	.8174	.8610	.8967	.9250	.9469	.9633
24	.4688	.5433	.6156	.6833	.7449	.7991	.8454	.8838	.9147	.9389
25	.3722	.4450	.5189	.5914	.6603	.7236	.7802	.8291	.8702	.9037
26	.2836	.3506	.4217	.4948	.5673	.6371	.7020	.7607	.8121	.8559
27	.2070	.2651	.3297	.3991	.4711	.5434	.6138	.6801	.7407	.7945
28	.1446	.1921	.2474	.3096	.3770	.4478	.5197	.5904	.6579	.7203
29	.0966	.1333	.1780	.2305	.2902	.3556	.4249	.4962	.5671	.6354
30	.0616	.0884	.1225	.1646	.2144	.2715	.3348	.4026	.4730	.5438
31	.0375	.0560	.0807	.1124	.1519	.1990	.2536	.3146	.3807	.4501
32	.0217	.0338	.0507	.0735	.1030	.1398	.1844	.2364	.2951	.3594
33	.0120	.0195	.0305	.0459	.0668	.0941	.1285	.1704	.2199	.2763
34	.0063	.0107	.0174	.0273	.0414	.0605	.0857	.1178	.1572	.2041
35	.0032	.0056	.0095	.0155	.0245	.0372	.0547	.0779	.1077	.1446
36	.0015	.0028	.0050	.0084	.0138	.0218	.0333	.0492	.0706	.0981
37	.0007	.0013	.0025	.0044	.0674	.0122	.0194	.0297	.0442	.0637
38	.0003	.0006	.0012	.0021	.0038	.0065	.0108	.0172	.0265	.0396
39	.0001	.0003	.0005	.0010	.0019	.0033	.0057	.0094	.0151	.0235
40	.0000	.0001	.0002	.0004	.0009	.0016	.0029	.0049	.0082	.0133
41	.0000	.0000	.0001	.0002	.0004	.0007	.0014	.0025	.0043	.0072
42	.0000	.0000	.0000	.0001	.0002	.0003	.0006	.0012	.0021	.0037
43	.0000	.0000	.0000	.0000	.0001	.0001	.0003	.0005	.0010	.0018
44	.0000	.0000	.0000	.0000	.0000	.0001	.0001	.0002	.0004	.0008
45	.0000	.0000	.0000	.0000	.0000	.0000	.0000	.0001	.0002	.0004
46	.0000	.0000	.0000	.0000	.0000	.0000	.0000	.0000	.0001	.0002
47	.0000	.0000	.0000	.0000	.0000	.0000	.0000	.0000	.0000	.0001

r \ p	.41	.42	.43	.44	.45	.46	.47	.48	.49	.50
1	1.0000	1.0000	1.0000	1.0000	1.0000	1.0000	1.0000	1.0000	1.0000	1.0000
2	1.0000	1.0000	1.0000	1.0000	1.0000	1.0000	1.0000	1.0000	1.0000	1.0000
3	1.0000	1.0000	1.0000	1.0000	1.0000	1.0000	1.0000	1.0000	1.0000	1.0000
4	1.0000	1.0000	1.0000	1.0000	1.0000	1.0000	1.0000	1.0000	1.0000	1.0000
5	1.0000	1.0000	1.0000	1.0000	1.0000	1.0000	1.0000	1.0000	1.0000	1.0000

Table A.1—*Continued*

n = 75 (Continued)

r \ p	.41	.42	.43	.44	.45	.46	.47	.48	.49	.50
6	1.0000	1.0000	1.0000	1.0000	1.0000	1.0000	1.0000	1.0000	1.0000	1.0000
7	1.0000	1.0000	1.0000	1.0000	1.0000	1.0000	1.0000	1.0000	1.0000	1.0000
8	1.0000	1.0000	1.0000	1.0000	1.0000	1.0000	1.0000	1.0000	1.0000	1.0000
9	1.0000	1.0000	1.0000	1.0000	1.0000	1.0000	1.0000	1.0000	1.0000	1.0000
10	1.0000	1.0000	1.0000	1.0000	1.0000	1.0000	1.0000	1.0000	1.0000	1.0000
11	1.0000	1.0000	1.0000	1.0000	1.0000	1.0000	1.0000	1.0000	1.0000	1.0000
12	1.0000	1.0000	1.0000	1.0000	1.0000	1.0000	1.0000	1.0000	1.0000	1.0000
13	1.0000	1.0000	1.0000	1.0000	1.0000	1.0000	1.0000	1.0000	1.0000	1.0000
14	1.0000	1.0000	1.0000	1.0000	1.0000	1.0000	1.0000	1.0000	1.0000	1.0000
15	1.0000	1.0000	1.0000	1.0000	1.0000	1.0000	1.0000	1.0000	1.0000	1.0000
16	.9999	1.0000	1.0000	1.0000	1.0000	1.0000	1.0000	1.0000	1.0000	1.0000
17	.9997	.9999	.9999	1.0000	1.0000	1.0000	1.0000	1.0000	1.0000	1.0000
18	.9993	.9996	.9998	.9999	1.0000	1.0000	1.0000	1.0000	1.0000	1.0000
19	.9985	.9991	.9995	.9997	.9999	.9999	1.0000	1.0000	1.0000	1.0000
20	.9966	.9980	.9989	.9994	.9996	.9998	.9999	1.0000	1.0000	1.0000
21	.9930	.9957	.9974	.9985	.9992	.9995	.9998	.9999	.9999	1.0000
22	.9864	.9914	.9947	.9968	.9981	.9989	.9994	.9997	.9998	.9999
23	.9752	.9837	.9895	.9935	.9960	.9976	.9986	.9992	.9996	.9998
24	.9573	.9709	.9806	.9874	.9920	.9951	.9971	.9983	.9990	.9995
25	.9303	.9507	.9660	.9771	.9850	.9904	.9940	.9964	.9979	.9988
26	.8920	.9210	.9436	.9607	.9733	.9823	.9886	.9928	.9956	.9974
27	.8408	.8795	.9110	.9358	.9548	.9690	.9793	.9865	.9914	.9947
28	.7763	.8251	.8664	.9003	.9274	.9484	.9643	.9759	.9841	.9899
29	.6994	.7575	.8087	.8525	.8890	.9184	.9415	.9591	.9721	.9815
30	.6128	.6782	.7382	.7917	.8380	.8770	.9088	.9340	.9535	.9680
31	.5206	.5901	.6566	.7184	.7741	.8228	.8643	.8985	.9259	.9473
32	.4275	.4975	.5673	.6348	.6981	.7559	.8070	.8509	.8876	.9173
33	.3386	.4053	.4746	.5444	.6127	.6774	.7371	.7905	.8369	.8760
34	.2581	.3183	.3835	.4519	.5216	.5904	.6564	.7178	.7734	.8222
35	.1889	.2405	.2986	.3622	.4295	.4988	.5679	.6349	.6980	.7557
36	.1327	.1745	.2236	.2795	.3413	.4073	.4760	.5453	.6132	.6778
37	.0892	.1214	.1608	.2074	.2610	.3208	.3855	.4534	.5227	.5912
38	.0574	.0809	.1108	.1477	.1918	.2431	.3009	.3640	.4310	.5000
39	.0353	.0515	.0730	.1007	.1352	.1770	.2259	.2815	.3429	.4088
40	.0207	.0313	.0460	.0657	.0913	.1235	.1627	.2092	.2626	.3222
41	.0116	.0182	.0277	.0410	.0590	.0825	.1123	.1492	.1932	.2443
42	.0062	.0101	.0159	.0244	.0364	.0527	.0742	.1018	.1363	.1778
43	.0031	.0053	.0087	.0139	.0214	.0321	.0468	.0665	.0920	.1240
44	.0015	.0027	.0045	.0075	.0120	.0186	.0282	.0414	.0593	.0827
45	.0007	.0013	.0022	.0039	.0064	.0103	.0162	.0246	.0365	.0527
46	.0003	.0006	.0011	.0019	.0032	.0054	.0088	.0139	.0214	.0320
47	.0001	.0002	.0005	.0009	.0016	.0027	.0046	.0075	.0119	.0185
48	.0000	.0001	.0002	.0004	.0007	.0013	.0023	.0038	.0063	.0101
49	.0000	.0000	.0001	.0002	.0003	.0006	.0011	.0019	.0032	.0053
50	.0000	.0000	.0000	.0001	.0001	.0002	.0005	.0009	.0015	.0026
51	.0000	.0000	.0000	.0000	.0000	.0001	.0002	.0004	.0007	.0012
52	.0000	.0000	.0000	.0000	.0000	.0000	.0001	.0002	.0003	.0005
53	.0000	.0000	.0000	.0000	.0000	.0000	.0000	.0001	.0001	.0002
54	.0000	.0000	.0000	.0000	.0000	.0000	.0000	.0000	.0000	.0001

n = 100

r \ p	.01	.02	.03	.04	.05	.06	.07	.08	.09	.10
1	.6340	.8674	.9524	.9831	.9941	.9979	.9993	.9998	.9999	1.0000
2	.2642	.5967	.8054	.9128	.9629	.9848	.9940	.9977	.9991	.9997
3	.0794	.3233	.5802	.7679	.8817	.9434	.9742	.9887	.9952	.9981
4	.0184	.1410	.3528	.5705	.7422	.8570	.9256	.9633	.9827	.9922
5	.0034	.0508	.1821	.3711	.5640	.7232	.8368	.9097	.9526	.9763
6	.0005	.0155	.0808	.2116	.3840	.5593	.7086	.8201	.8955	.9424
7	.0001	.0041	.0312	.1064	.2340	.3936	.5557	.6968	.8060	.8828
8	.0000	.0009	.0106	.0475	.1280	.2517	.4012	.5529	.6872	.7939
9	.0000	.0002	.0032	.0190	.0631	.1463	.2660	.4074	.5506	.6791
10	.0000	.0000	.0009	.0068	.0282	.0775	.1620	.2780	.4125	.5487
11	.0000	.0000	.0002	.0022	.0115	.0376	.0908	.1757	.2882	.4168
12	.0000	.0000	.0000	.0007	.0043	.0168	.0469	.1028	.1876	.2970
13	.0000	.0000	.0000	.0002	.0015	.0069	.0224	.0559	.1138	.1982
14	.0000	.0000	.0000	.0000	.0005	.0026	.0099	.0282	.0645	.1239
15	.0000	.0000	.0000	.0000	.0001	.0009	.0041	.0133	.0341	.0726

Table A.1—*Continued*

n = 100 (Continued)

r \ p	.01	.02	.03	.04	.05	.06	.07	.08	.09	.10
16	.0000	.0000	.0000	.0000	.0000	.0003	.0016	.0058	.0169	.0399
17	.0000	.0000	.0000	.0000	.0000	.0001	.0006	.0024	.0078	.0206
18	.0000	.0000	.0000	.0000	.0000	.0000	.0002	.0009	.0034	.0100
19	.0000	.0000	.0000	.0000	.0000	.0000	.0001	.0003	.0014	.0046
20	.0000	.0000	.0000	.0000	.0000	.0000	.0000	.0001	.0005	.0020
21	.0000	.0000	.0000	.0000	.0000	.0000	.0000	.0000	.0002	.0008
22	.0000	.0000	.0000	.0000	.0000	.0000	.0000	.0000	.0001	.0003
23	.0000	.0000	.0000	.0000	.0000	.0000	.0000	.0000	.0000	.0001

r \ p	.11	.12	.13	.14	.15	.16	.17	.18	.19	.20
1	1.0000	1.0000	1.0000	1.0000	1.0000	1.0000	1.0000	1.0000	1.0000	1.0000
2	.9999	1.0000	1.0000	1.0000	1.0000	1.0000	1.0000	1.0000	1.0000	1.0000
3	.9992	.9997	.9999	1.0000	1.0000	1.0000	1.0000	1.0000	1.0000	1.0000
4	.9966	.9985	.9994	.9998	.9999	1.0000	1.0000	1.0000	1.0000	1.0000
5	.9886	.9947	.9977	.9990	.9996	.9998	.9999	1.0000	1.0000	1.0000
6	.9698	.9848	.9926	.9966	.9984	.9993	.9997	.9999	1.0000	1.0000
7	.9328	.9633	.9808	.9903	.9953	.9978	.9990	.9996	.9998	.9999
8	.8715	.9239	.9569	.9766	.9878	.9939	.9970	.9986	.9994	.9997
9	.7835	.8614	.9155	.9508	.9725	.9853	.9924	.9962	.9982	.9991
10	.6722	.7743	.8523	.9078	.9449	.9684	.9826	.9908	.9953	.9977
11	.5471	.6663	.7663	.8440	.9006	.9393	.9644	.9800	.9891	.9943
12	.4206	.5458	.6611	.7591	.8365	.8939	.9340	.9605	.9773	.9874
13	.3046	.4239	.5446	.6566	.7527	.8297	.8876	.9289	.9567	.9747
14	.2076	.3114	.4268	.5436	.6526	.7469	.8234	.8819	.9241	.9531
15	.1330	.2160	.3173	.4294	.5428	.6490	.7417	.8177	.8765	.9196
16	.0802	.1414	.2236	.3227	.4317	.5420	.6458	.7370	.8125	.8715
17	.0456	.0874	.1492	.2305	.3275	.4338	.5414	.6429	.7327	.8077
18	.0244	.0511	.0942	.1563	.2367	.3319	.4357	.5408	.6403	.7288
19	.0123	.0282	.0564	.1006	.1628	.2424	.3359	.4374	.5403	.6379
20	.0059	.0147	.0319	.0614	.1065	.1689	.2477	.3395	.4391	.5398
21	.0026	.0073	.0172	.0356	.0663	.1121	.1745	.2525	.3429	.4405
22	.0011	.0034	.0088	.0196	.0393	.0710	.1174	.1797	.2570	.3460
23	.0005	.0015	.0042	.0103	.0221	.0428	.0754	.1223	.1846	.2611
24	.0002	.0006	.0020	.0051	.0119	.0246	.0462	.0796	.1270	.1891
25	.0001	.0003	.0009	.0024	.0061	.0135	.0271	.0496	.0837	.1314
26	.0000	.0001	.0004	.0011	.0030	.0071	.0151	.0295	.0528	.0875
27	.0000	.0000	.0001	.0005	.0014	.0035	.0081	.0168	.0318	.0558
28	.0000	.0000	.0001	.0002	.0006	.0017	.0041	.0091	.0184	.0342
29	.0000	.0000	.0000	.0001	.0003	.0008	.0020	.0048	.0102	.0200
30	.0000	.0000	.0000	.0000	.0001	.0003	.0009	.0024	.0054	.0112
31	.0000	.0000	.0000	.0000	.0000	.0001	.0004	.0011	.0027	.0061
32	.0000	.0000	.0000	.0000	.0000	.0001	.0002	.0005	.0013	.0031
33	.0000	.0000	.0000	.0000	.0000	.0000	.0001	.0002	.0006	.0016
34	.0000	.0000	.0000	.0000	.0000	.0000	.0000	.0001	.0003	.0007
35	.0000	.0000	.0000	.0000	.0000	.0000	.0000	.0000	.0001	.0003
36	.0000	.0000	.0000	.0000	.0000	.0000	.0000	.0000	.0000	.0001
37	.0000	.0000	.0000	.0000	.0000	.0000	.0000	.0000	.0000	.0001

r \ p	.21	.22	.23	.24	.25	.26	.27	.28	.29	.30
1	1.0000	1.0000	1.0000	1.0000	1.0000	1.0000	1.0000	1.0000	1.0000	1.0000
2	1.0000	1.0000	1.0000	1.0000	1.0000	1.0000	1.0000	1.0000	1.0000	1.0000
3	1.0000	1.0000	1.0000	1.0000	1.0000	1.0000	1.0000	1.0000	1.0000	1.0000
4	1.0000	1.0000	1.0000	1.0000	1.0000	1.0000	1.0000	1.0000	1.0000	1.0000
5	1.0000	1.0000	1.0000	1.0000	1.0000	1.0000	1.0000	1.0000	1.0000	1.0000
6	1.0000	1.0000	1.0000	1.0000	1.0000	1.0000	1.0000	1.0000	1.0000	1.0000
7	1.0000	1.0000	1.0000	1.0000	1.0000	1.0000	1.0000	1.0000	1.0000	1.0000
8	.9999	1.0000	1.0000	1.0000	1.0000	1.0000	1.0000	1.0000	1.0000	1.0000
9	.9996	1.9998	1.9999	1.0000	1.0000	1.0000	1.0000	1.0000	1.0000	1.0000
10	.9989	.9995	.9998	.9999	1.0000	1.0000	1.0000	1.0000	1.0000	1.0000
11	.9971	.9986	.9993	.9997	.9999	.9999	1.0000	1.0000	1.0000	1.0000
12	.9933	.9965	.9983	.9992	.9996	.9998	.9999	1.0000	1.0000	1.0000
13	.9857	.9922	.9959	.9979	.9990	.9995	.9998	.9999	1.0000	1.0000
14	.9721	.9840	.9911	.9953	.9975	.9988	.9994	.9997	.9999	.9999
15	.9496	.9695	.9823	.9900	.9946	.9972	.9986	.9993	.9997	.9998
16	.9153	.9462	.9671	.9806	.9889	.9939	.9967	.9983	.9992	.9996
17	.8668	.9112	.9430	.9647	.9789	.9878	.9932	.9963	.9981	.9990
18	.8032	.8625	.9074	.9399	.9624	.9773	.9867	.9925	.9959	.9978
19	.7252	.7991	.8585	.9038	.9370	.9601	.9757	.9856	.9918	.9955
20	.6358	.7220	.7953	.8547	.9005	.9342	.9580	.9741	.9846	.9911

Table A.1—*Continued*

n = 100 (Continued)

r \ p	.21	.22	.23	.24	.25	.26	.27	.28	.29	.30
21	.5394	.6338	.7189	.7918	.8512	.8973	.9316	.9560	.9726	.9835
22	.4419	.5391	.6320	.7162	.7886	.8479	.8943	.9291	.9540	.9712
23	.3488	.4432	.5388	.6304	.7136	.7856	.8448	.8915	.9267	.9521
24	.2649	.3514	.4444	.5386	.6289	.7113	.7828	.8420	.8889	.9245
25	.1933	.2684	.3539	.4455	.5383	.6276	.7091	.7802	.8393	.8864
26	.1355	.1972	.2717	.3561	.4465	.5381	.6263	.7071	.7778	.8369
27	.0911	.1393	.2009	.2748	.3583	.4475	.5380	.6252	.7053	.7756
28	.0588	.0945	.1429	.2043	.2776	.3602	.4484	.5378	.6242	.7036
29	.0364	.0616	.0978	.1463	.2075	.2803	.3621	.4493	.5377	.6232
30	.0216	.0386	.0643	.1009	.1495	.2105	.2828	.3638	.4501	.5377
31	.0123	.0232	.0406	.0669	.1038	.1526	.2134	.2851	.3654	.4509
32	.0067	.0134	.0247	.0427	.0693	.1065	.1554	.2160	.2873	.3669
33	.0035	.0074	.0144	.0262	.0446	.0717	.1091	.1580	.2184	.2893
34	.0018	.0039	.0081	.0154	.0276	.0465	.0739	.1116	.1605	.2207
35	.0009	.0020	.0044	.0087	.0164	.0290	.0482	.0760	.1139	.1629
36	.0004	.0010	.0023	.0048	.0094	.0174	.0303	.0499	.0780	.1161
37	.0002	.0005	.0011	.0025	.0052	.0101	.0183	.0316	.0515	.0799
38	.0001	.0002	.0005	.0013	.0027	.0056	.0107	.0193	.0328	.0530
39	.0000	.0001	.0002	.0006	.0014	.0030	.0060	.0113	.0201	.0340
40	.0000	.0000	.0001	.0003	.0007	.0015	.0032	.0064	.0119	.0210
41	.0000	.0000	.0000	.0001	.0003	.0008	.0017	.0035	.0068	.0125
42	.0000	.0000	.0000	.0001	.0001	.0004	.0008	.0018	.0037	.0072
43	.0000	.0000	.0000	.0000	.0001	.0002	.0004	.0009	.0020	.0040
44	.0000	.0000	.0000	.0000	.0000	.0001	.0002	.0005	.0010	.0021
45	.0000	.0000	.0000	.0000	.0000	.0000	.0001	.0002	.0005	.0011
46	.0000	.0000	.0000	.0000	.0000	.0000	.0000	.0001	.0002	.0005
47	.0000	.0000	.0000	.0000	.0000	.0000	.0000	.0000	.0001	.0003
48	.0000	.0000	.0000	.0000	.0000	.0000	.0000	.0000	.0000	.0001
49	.0000	.0000	.0000	.0000	.0000	.0000	.0000	.0000	.0000	.0001

r \ p	.31	.32	.33	.34	.35	.36	.37	.38	.39	.40
1	1.0000	1.0000	1.0000	1.0000	1.0000	1.0000	1.0000	1.0000	1.0000	1.0000
2	1.0000	1.0000	1.0000	1.0000	1.0000	1.0000	1.0000	1.0000	1.0000	1.0000
3	1.0000	1.0000	1.0000	1.0000	1.0000	1.0000	1.0000	1.0000	1.0000	1.0000
4	1.0000	1.0000	1.0000	1.0000	1.0000	1.0000	1.0000	1.0000	1.0000	1.0000
5	1.0000	1.0000	1.0000	1.0000	1.0000	1.0000	1.0000	1.0000	1.0000	1.0000
6	1.0000	1.0000	1.0000	1.0000	1.0000	1.0000	1.0000	1.0000	1.0000	1.0000
7	1.0000	1.0000	1.0000	1.0000	1.0000	1.0000	1.0000	1.0000	1.0000	1.0000
8	1.0000	1.0000	1.0000	1.0000	1.0000	1.0000	1.0000	1.0000	1.0000	1.0000
9	1.0000	1.0000	1.0000	1.0000	1.0000	1.0000	1.0000	1.0000	1.0000	1.0000
10	1.0000	1.0000	1.0000	1.0000	1.0000	1.0000	1.0000	1.0000	1.0000	1.0000
11	1.0000	1.0000	1.0000	1.0000	1.0000	1.0000	1.0000	1.0000	1.0000	1.0000
12	1.0000	1.0000	1.0000	1.0000	1.0000	1.0000	1.0000	1.0000	1.0000	1.0000
13	1.0000	1.0000	1.0000	1.0000	1.0000	1.0000	1.0000	1.0000	1.0000	1.0000
14	1.0000	1.0000	1.0000	1.0000	1.0000	1.0000	1.0000	1.0000	1.0000	1.0000
15	.9999	1.0000	1.0000	1.0000	1.0000	1.0000	1.0000	1.0000	1.0000	1.0000
16	.9998	.9999	1.0000	1.0000	1.0000	1.0000	1.0000	1.0000	1.0000	1.0000
17	.9995	.9998	.9999	1.0000	1.0000	1.0000	1.0000	1.0000	1.0000	1.0000
18	.9989	.9995	.9997	.9999	.9999	1.0000	1.0000	1.0000	1.0000	1.0000
19	.9976	.9988	.9994	.9997	.9999	.9999	1.0000	1.0000	1.0000	1.0000
20	.9950	.9973	.9986	.9993	.9997	.9998	.9999	1.0000	1.0000	1.0000
21	.9904	.9946	.9971	.9985	.9992	.9996	.9998	.9999	1.0000	1.0000
22	.9825	.9898	.9942	.9968	.9983	.9991	.9996	.9998	.9999	1.0000
23	.9698	.9816	.9891	.9938	.9966	.9982	.9991	.9995	.9998	.9999
24	.9504	.9685	.9806	.9885	.9934	.9963	.9980	.9990	.9995	.9997
25	.9224	.9487	.9672	.9797	.9879	.9930	.9961	.9979	.9989	.9994
26	.8841	.9204	.9471	.9660	.9789	.9873	.9926	.9958	.9977	.9988
27	.8346	.8820	.9185	.9456	.9649	.9780	.9867	.9922	.9956	.9976
28	.7736	.8325	.8800	.9168	.9442	.9638	.9773	.9862	.9919	.9954
29	.7021	.7717	.8305	.8781	.9152	.9429	.9628	.9765	.9857	.9916
30	.6224	.7007	.7699	.8287	.8764	.9137	.9417	.9618	.9759	.9852
31	.5376	.6216	.6994	.7684	.8270	.8748	.9123	.9405	.9610	.9752
32	.4516	.5376	.6209	.6982	.7669	.8254	.8733	.9110	.9395	.9602
33	.3683	.4523	.5375	.6203	.6971	.7656	.8240	.8720	.9098	.9385
34	.2912	.3696	.4530	.5375	.6197	.6961	.7643	.8227	.8708	.9087
35	.2229	.2929	.3708	.4536	.5376	.6192	.6953	.7632	.8216	.8697
36	.1650	.2249	.2946	.3720	.4542	.5376	.6188	.6945	.7623	.8205
37	.1181	.1671	.2268	.2961	.3731	.4547	.5377	.6184	.6938	.7614
38	.0816	.1200	.1690	.2285	.2976	.3741	.4553	.5377	.6181	.6932
39	.0545	.0833	.1218	.1708	.2301	.2989	.3750	.4558	.5378	.6178
49	.0351	.0558	.0849	.1235	.1724	.2316	.3001	.3759	.4562	.5379

Table A.1—*Continued*

n = 100 (Continued)

r \ p	.31	.32	.33	.34	.35	.36	.37	.38	.39	.40
41	.0218	.0361	.0571	.0863	.1250	.1739	.2330	.3012	.3767	.4567
42	.0131	.0226	.0371	.0583	.0877	.1265	.1753	.2343	.3023	.3775
43	.0075	.0136	.0233	.0380	.0594	.0889	.1278	.1766	.2355	.3033
44	.0042	.0079	.0141	.0240	.0389	.0605	.0901	.1290	.1778	.2365
45	.0023	.0044	.0082	.0146	.0246	.0397	.0614	.0911	.1301	.1789
46	.0012	.0024	.0046	.0085	.0150	.0252	.0405	.0623	.0921	.1311
47	.0006	.0012	.0025	.0048	.0088	.0154	.0257	.0411	.0631	.0930
48	.0003	.0006	.0013	.0026	.0050	.0091	.0158	.0262	.0417	.0638
49	.0001	.0003	.0007	.0014	.0027	.0052	.0094	.0162	.0267	.0423
50	.0001	.0001	.0003	.0007	.0015	.0029	.0054	.0096	.0165	.0271
51	.0000	.0001	.0002	.0003	.0007	.0015	.0030	.0055	.0098	.0168
52	.0000	.0000	.0001	.0002	.0004	.0008	.0016	.0030	.0056	.0100
53	.0000	.0000	.0000	.0001	.0002	.0004	.0008	.0016	.0031	.0058
54	.0000	.0000	.0000	.0000	.0001	.0002	.0004	.0008	.0017	.0032
55	.0000	.0000	.0000	.0000	.0000	.0001	.0002	.0004	.0009	.0017
56	.0000	.0000	.0000	.0000	.0000	.0000	.0001	.0002	.0004	.0009
57	.0000	.0000	.0000	.0000	.0000	.0000	.0000	.0001	.0002	.0004
58	.0000	.0000	.0000	.0000	.0000	.0000	.0000	.0000	.0001	.0002
59	.0000	.0000	.0000	.0000	.0000	.0000	.0000	.0000	.0000	.0001

r \ p	.41	.42	.43	.44	.45	.46	.47	.48	.49	.50
1	1.0000	1.0000	1.0000	1.0000	1.0000	1.0000	1.0000	1.0000	1.0000	1.0000
2	1.0000	1.0000	1.0000	1.0000	1.0000	1.0000	1.0000	1.0000	1.0000	1.0000
3	1.0000	1.0000	1.0000	1.0000	1.0000	1.0000	1.0000	1.0000	1.0000	1.0000
4	1.0000	1.0000	1.0000	1.0000	1.0000	1.0000	1.0000	1.0000	1.0000	1.0000
5	1.0000	1.0000	1.0000	1.0000	1.0000	1.0000	1.0000	1.0000	1.0000	1.0000
6	1.0000	1.0000	1.0000	1.0000	1.0000	1.0000	1.0000	1.0000	1.0000	1.0000
7	1.0000	1.0000	1.0000	1.0000	1.0000	1.0000	1.0000	1.0000	1.0000	1.0000
8	1.0000	1.0000	1.0000	1.0000	1.0000	1.0000	1.0000	1.0000	1.0000	1.0000
9	1.0000	1.0000	1.0000	1.0000	1.0000	1.0000	1.0000	1.0000	1.0000	1.0000
10	1.0000	1.0000	1.0000	1.0000	1.0000	1.0000	1.0000	1.0000	1.0000	1.0000
11	1.0000	1.0000	1.0000	1.0000	1.0000	1.0000	1.0000	1.0000	1.0000	1.0000
12	1.0000	1.0000	1.0000	1.0000	1.0000	1.0000	1.0000	1.0000	1.0000	1.0000
13	1.0000	1.0000	1.0000	1.0000	1.0000	1.0000	1.0000	1.0000	1.0000	1.0000
14	1.0000	1.0000	1.0000	1.0000	1.0000	1.0000	1.0000	1.0000	1.0000	1.0000
15	1.0000	1.0000	1.0000	1.0000	1.0000	1.0000	1.0000	1.0000	1.0000	1.0000
16	1.0000	1.0000	1.0000	1.0000	1.0000	1.0000	1.0000	1.0000	1.0000	1.0000
17	1.0000	1.0000	1.0000	1.0000	1.0000	1.0000	1.0000	1.0000	1.0000	1.0000
18	1.0000	1.0000	1.0000	1.0000	1.0000	1.0000	1.0000	1.0000	1.0000	1.0000
19	1.0000	1.0000	1.0000	1.0000	1.0000	1.0000	1.0000	1.0000	1.0000	1.0000
20	1.0000	1.0000	1.0000	1.0000	1.0000	1.0000	1.0000	1.0000	1.0000	1.0000
21	1.0000	1.0000	1.0000	1.0000	1.0000	1.0000	1.0000	1.0000	1.0000	1.0000
22	1.0000	1.0000	1.0000	1.0000	1.0000	1.0000	1.0000	1.0000	1.0000	1.0000
23	1.0000	1.0000	1.0000	1.0000	1.0000	1.0000	1.0000	1.0000	1.0000	1.0000
24	.9999	.9999	1.0000	1.0000	1.0000	1.0000	1.0000	1.0000	1.0000	1.0000
25	.9997	.9999	.9999	1.0000	1.0000	1.0000	1.0000	1.0000	1.0000	1.0000
26	.9994	.9997	.9999	.9999	1.0000	1.0000	1.0000	1.0000	1.0000	1.0000
27	.9987	.9994	.9997	.9998	.9999	1.0000	1.0000	1.0000	1.0000	1.0000
28	.9975	.9987	.9993	.9997	.9998	.9999	1.0000	1.0000	1.0000	1.0000
29	.9952	.9974	.9986	.9993	.9996	.9998	.9999	1.0000	1.0000	1.0000
30	.9913	.9950	.9972	.9985	.9992	.9996	.9998	.9999	1.0000	1.0000
31	.9848	.9910	.9948	.9971	.9985	.9992	.9996	.9998	.9999	1.0000
32	.9746	.9844	.9907	.9947	.9970	.9984	.9992	.9996	.9998	.9999
33	.9594	.9741	.9840	.9905	.9945	.9969	.9984	.9991	.9996	.9998
34	.9376	.9587	.9736	.9837	.9902	.9944	.9969	.9983	.9991	.9996
35	.9078	.9368	.9581	.9732	.9834	.9900	.9942	.9968	.9983	.9991
36	.8687	.9069	.9361	.9576	.9728	.9831	.9899	.9941	.9967	.9982
37	.8196	.8678	.9061	.9355	.9571	.9724	.9829	.9897	.9941	.9967
38	.7606	.8188	.8670	.9054	.9349	.9567	.9721	.9827	.9896	.9940
39	.6927	.7599	.8181	.8663	.9049	.9345	.9563	.9719	.9825	.9895
40	.6176	.6922	.7594	.8174	.8657	.9044	.9341	.9561	.9717	.9824
41	.5380	.6174	.6919	.7589	.8169	.8653	.9040	.9338	.9558	.9716
42	.4571	.5382	.6173	.6916	.7585	.8165	.8649	.9037	.9335	.9557
43	.3782	.4576	.5383	.6173	.6913	.7582	.8162	.8646	.9035	.9334
44	.3041	.3788	.4580	.5385	.6172	.6912	.7580	.8160	.8645	.9033
45	.2375	.3049	.3794	.4583	.5387	.6173	.6911	.7579	.8159	.8644

Table A.1—*Continued*

	n = 100 (Continued)									
r \ p	.41	.42	.43	.44	.45	.46	.47	.48	.49	.50
46	.1799	.2384	.3057	.3799	.4587	.5389	.6173	.6911	.7579	.8159
47	.1320	.1807	.2391	.3063	.3804	.4590	.5391	.6174	.6912	.7579
48	.0938	.1328	.1815	.2398	.3069	.3809	.4593	.5393	.6176	.6914
49	.0644	.0944	.1335	.1822	.2404	.3074	.3813	.4596	.5395	.6178
50	.0428	.0650	.0950	.1341	.1827	.2409	.3078	.3816	.4599	.5398
51	.0275	.0432	.0655	.0955	.1346	.1832	.2413	.3082	.3819	.4602
52	.0170	.0278	.0436	.0659	.0960	.1350	.1836	.2417	.3084	.3822
53	.0102	.0172	.0280	.0439	.0662	0963	.1353	.1838	.2419	.3086
54	.0059	.0103	.0174	.0282	.0441	.0664	.0965	.1355	.1840	.2421
55	.0033	.0059	.0104	.0175	.0284	.0443	.0666	.0967	.1356	.1841
56	.0017	.0033	.0060	.0105	.0176	.0285	.0444	.0667	.0967	.1356
57	.0009	.0018	.0034	.0061	.0106	.0177	.0286	.0444	.0667	.0967
58	.0004	.0009	.0018	.0034	.0061	.0106	.0177	.0286	.0444	.0666
59	.0002	.0005	.0009	.0018	.0034	.0061	.0106	.0177	.0285	.0443
60	.0001	.0002	.0005	.0009	.0018	.0034	.0061	.0106	.0177	.0284
61	.0000	.0001	.0002	.0005	.0009	.0018	.0034	.0061	.0106	.0176
62	.0000	.0000	.0001	.0002	.0005	.0009	.0018	.0034	.0061	.0105
63	.0000	.0000	.0000	.0001	.0002	.0005	.0009	.0018	.0034	.0060
64	.0000	.0000	.0000	.0000	.0001	.0002	.0005	.0009	.0018	.0033
65	.0000	.0000	.0000	.0000	.0000	.0001	.0002	.0005	.0009	.0018
66	.0000	.0000	.0000	.0000	.0000	.0000	.0001	.0002	.0004	.0009
67	.0000	.0000	.0000	.0000	.0000	.0000	.0000	.0001	.0002	.0004
68	.0000	.0000	.0000	.0000	.0000	.0000	.0000	.0000	.0001	.0002
69	.0000	.0000	.0000	.0000	.0000	.0000	.0000	.0000	.0000	.0001

Table A.2
Areas for the Standard Normal Probability Distribution

z	.00	.01	.02	.03	.04	.05	.06	.07	.08	.09
0.0	.0000	.0040	.0080	.0120	.0160	.0199	.0239	.0279	.0319	.0359
0.1	.0398	.0438	.0478	.0517	.0557	.0596	.0636	.0675	.0714	.0753
0.2	.0793	.0832	.0871	.0910	.0948	.0987	.1026	.1064	.1103	.1141
0.3	.1179	.1217	.1255	.1293	.1331	.1368	.1406	.1443	.1480	.1517
0.4	.1554	.1591	.1628	.1664	.1700	.1736	.1772	.1808	.1844	.1879
0.5	.1915	.1950	.1985	.2019	.2054	.2088	.2123	.2157	.2190	.2224
0.6	.2257	.2291	.2324	.2357	.2389	.2422	.2454	.2486	.2518	.2549
0.7	.2580	.2612	.2642	.2673	.2704	.2734	.2764	.2794	.2823	.2852
0.8	.2881	.2910	.2939	.2967	.2995	.3023	.3051	.3078	.3106	.3133
0.9	.3159	.3186	.3212	.3238	.3264	.3289	.3315	.3340	.3365	.3389
1.0	.3413	.3438	.3461	.3485	.3508	.3531	.3554	.3577	.3599	.3621
1.1	.3643	.3665	.3686	.3708	.3729	.3749	.3770	.3790	.3810	.3830
1.2	.3849	.3869	.3888	.3907	.3925	.3944	.3962	.3980	.3997	.4015
1.3	.4032	.4049	.4066	.4082	.4099	.4115	.4131	.4147	.4162	.4177
1.4	.4192	.4207	.4222	.4236	.4251	.4265	.4279	.4292	.4306	.4319
1.5	.4332	.4345	.4357	.4370	.4382	.4394	.4406	.4418	.4429	.4441
1.6	.4452	.4463	.4474	.4484	.4495	.4505	.4515	.4525	.4535	.4545
1.7	.4554	.4564	.4573	.4582	.4591	.4599	.4608	.4616	.4625	.4633
1.8	.4641	.4649	.4656	.4664	.4671	.4678	.4686	.4693	.4699	.4706
1.9	.4713	.4719	.4726	.4732	.4738	.4744	.4750	.4756	.4761	.4767
2.0	.4772	.4778	.4783	.4788	.4793	.4798	.4803	.4808	.4812	.4817
2.1	.4821	.4826	.4830	.4834	.4838	.4842	.4846	.4850	.4854	.4857
2.2	.4861	.4864	.4868	.4871	.4875	.4878	.4881	.4884	.4887	.4890
2.3	.4893	.4896	.4898	.4901	.4904	.4906	.4909	.4911	.4913	.4916
2.4	.4918	.4920	.4922	.4925	.4927	.4929	.4931	.4932	.4934	.4936
2.5	.4938	.4940	.4941	.4943	.4945	.4946	.4948	.4949	.4951	.4952
2.6	.4953	.4955	.4956	.4957	.4959	.4960	.4961	.4962	.4963	.4964
2.7	.4965	.4966	.4967	.4968	.4969	.4970	.4971	.4972	.4973	.4974
2.8	.4974	.4975	.4976	.4977	.4977	.4978	.4979	.4970	.4980	.4981
2.9	.4981	.4982	.4982	.4983	.4984	.4984	.4985	.4985	.4986	.4986
3.0	.49865	.4987	.4987	.4988	.4988	.4989	.4989	.4989	.4990	.4990
4.0	.4999683									

Table A.3
Critical Values of the *t* Distribution, One-Tail Area

	Level of Significance, α				
df	.10	.05	.025	.010	.005
1	3.078	6.314	12.706	31.821	63.657
2	1.886	2.920	4.303	6.965	9.925
3	1.638	2.353	3.182	4.541	5.841
4	1.533	2.132	2.776	3.747	4.604
5	1.476	2.015	2.571	3.365	4.032
6	1.440	1.943	2.447	3.143	3.707
7	1.415	1.895	2.365	2.998	3.499
8	1.397	1.860	2.306	2.896	3.355
9	1.383	1.833	2.262	2.821	3.250
10	1.372	1.812	2.228	2.764	3.169
11	1.363	1.796	2.201	2.718	3.106
12	1.356	1.782	2.179	2.681	3.055
13	1.350	1.771	2.160	2.650	3.012
14	1.345	1.761	2.145	2.624	2.977
15	1.341	1.753	2.131	2.602	2.947
16	1.337	1.746	2.120	2.583	2.921
17	1.333	1.740	2.110	2.567	2.898
18	1.330	1.734	2.101	2.552	2.878
19	1.328	1.729	2.093	2.539	2.861
20	1.325	1.725	2.086	2.528	2.845
21	1.323	1.721	2.080	2.518	2.831
22	1.321	1.717	2.074	2.508	2.819
23	1.319	1.714	2.069	2.500	2.807
24	1.318	1.711	2.064	2.492	2.797
25	1.316	1.708	2.060	2.485	2.787
26	1.315	1.706	2.056	2.479	2.779
27	1.314	1.703	2.052	2.473	2.771
28	1.313	1.701	2.048	2.467	2.763
29	1.311	1.699	2.045	2.462	2.756
30	1.310	1.697	2.042	2.457	2.750
∞	1.28155	1.64485	1.95996	2.32634	2.57582

Table A.4
Critical Values of the Chi-Square Distribution

	Tail Area to the Right of χ^2							
df	.995	.090	.975	.950	.050	.025	.010	.005
1	--	--	--	.004	3.84	5.02	6.63	7.88
2	.01	.02	.05	.10	5.99	7.38	9.21	10.60
3	.07	.11	.22	.35	7.81	9.35	11.34	12.84
4	.21	.30	.48	.71	9.49	11.14	13.28	14.86
5	.41	.55	.83	1.15	11.07	12.83	15.09	16.75
6	.68	.87	1.24	1.64	12.59	14.45	16.81	18.55
7	.99	1.24	1.69	2.17	14.07	16.01	18.48	20.28
8	1.34	1.65	2.18	2.73	15.51	17.53	20.09	21.96
9	1.73	2.09	2.70	3.33	16.92	19.02	21.67	23.59
10	2.16	2.56	3.25	3.94	18.31	20.48	23.21	25.19
11	2.60	3.05	3.82	4.57	19.68	21.92	24.72	26.76
12	3.07	3.57	4.40	5.23	21.03	23.34	26.22	28.30
13	3.57	4.11	5.01	5.89	22.36	24.74	27.69	29.82
14	4.07	4.66	5.63	6.57	23.68	26.12	29.14	31.32
15	4.60	5.23	6.26	7.26	25.00	27.49	30.58	32.80
16	5.14	5.81	6.91	7.96	26.30	28.85	32.00	34.27
17	5.70	6.41	7.56	8.67	27.59	30.19	33.41	35.72
18	6.26	7.01	8.23	9.39	28.87	31.53	34.81	37.16
19	6.84	7.63	8.91	10.12	30.14	32.85	36.19	38.58
20	7.43	8.26	9.59	10.85	31.41	34.17	37.57	40.00
21	8.03	8.90	10.28	11.59	32.67	35.48	38.93	41.40
22	8.64	9.54	10.98	12.34	33.92	36.78	40.29	42.80
23	9.26	10.20	11.69	13.09	35.17	38.08	41.64	44.18
24	9.89	10.86	12.40	13.85	36.42	39.36	42.98	45.56
25	10.52	11.52	13.12	14.61	37.65	40.65	44.31	46.93
26	11.16	12.20	13.84	15.38	38.89	41.92	45.64	48.29
27	11.81	12.88	14.57	16.15	40.11	43.19	46.96	49.64
28	12.46	13.56	15.31	16.93	41.34	44.46	48.28	50.99
29	13.12	14.26	16.05	17.71	42.56	45.72	49.59	51.34
30	13.79	14.95	16.79	18.49	43.77	46.98	50.89	53.67
40	20.71	22.16	24.43	26.51	55.76	59.34	63.69	66.77
50	27.99	29.71	32.36	34.76	67.50	71.42	76.15	79.49
60	35.53	37.48	40.48	43.19	79.08	83.30	88.38	91.95
70	43.28	45.44	48.76	51.74	90.53	95.02	100.43	104.22
80	51.17	53.54	57.15	60.39	101.88	106.63	112.33	116.32
90	59.20	61.75	65.65	69.13	113.14	118.14	124.12	128.30
100	67.33	70.06	74.22	77.93	124.34	129.56	135.81	140.17

NOTE: Dashes indicate no values are calculated for these levels of significance.

Table A.5
Critical Values of the *F* Distribution

df	α = .05								
	1	2	3	4	5	6	7	8	9
1	161.4	199.5	215.7	224.6	230.2	234.0	236.8	238.9	240.5
2	18.51	19.00	19.16	19.25	19.30	19.33	19.35	19.37	19.38
3	10.13	9.55	9.28	9.12	9.01	8.94	8.89	8.85	8.81
4	7.71	6.94	6.59	6.39	6.26	6.16	6.09	6.04	6.00
5	6.61	5.79	5.41	5.19	5.05	4.95	4.88	4.82	4.77
6	5.99	5.14	4.76	4.53	4.39	4.28	4.21	4.15	4.10
7	5.59	4.74	4.35	4.12	3.97	3.87	3.79	3.73	3.68
8	5.32	4.46	4.07	3.84	3.69	3.58	3.50	3.44	3.39
9	5.12	4.26	3.86	3.63	3.48	3.37	3.29	3.23	3.18
10	4.96	4.10	3.71	3.48	3.33	3.22	3.14	3.07	3.02
11	4.84	3.98	3.59	3.36	3.20	3.09	3.01	2.95	2.90
12	4.75	3.89	3.49	3.26	3.11	3.00	2.91	2.85	2.80
13	4.67	3.81	3.41	3.18	3.03	2.92	2.83	2.77	2.71
14	4.60	3.74	3.34	3.11	2.96	2.85	2.76	2.70	2.65
15	4.54	3.68	3.29	3.06	2.90	2.79	2.71	2.64	2.59
16	4.49	3.63	3.24	3.01	2.85	2.74	2.66	2.59	2.54
17	4.45	3.59	3.20	2.96	2.81	2.70	2.61	2.55	2.49
18	4.41	3.55	3.16	2.93	2.77	2.66	2.58	2.51	2.46
19	4.38	3.52	3.13	2.90	2.74	2.63	2.54	2.48	2.42
20	4.35	3.49	3.10	2.87	2.71	2.60	2.51	2.45	2.39
21	4.32	3.47	3.07	2.84	2.68	2.57	2.49	2.42	2.37
22	4.30	3.44	3.05	2.82	2.66	2.55	2.46	2.40	2.34
23	4.28	3.42	3.03	2.80	2.64	2.53	2.44	2.37	2.32
24	4.26	3.40	3.01	2.78	2.62	2.51	2.42	2.36	2.30
25	4.24	3.39	2.99	2.76	2.60	2.49	2.40	2.34	2.28
26	4.23	3.37	2.98	2.74	2.59	2.47	2.39	2.32	2.27
27	4.21	3.35	2.96	2.73	2.57	2.46	2.37	2.31	2.25
28	4.20	3.34	2.95	2.71	2.56	2.45	2.36	2.29	2.24
29	4.18	3.33	2.93	2.70	2.55	2.43	2.35	2.28	2.22
30	4.17	3.32	2.92	2.69	2.53	2.42	2.33	2.27	2.21
40	4.08	3.23	2.84	2.61	2.45	2.34	2.25	2.18	2.12
60	4.00	3.15	2.76	2.53	2.37	2.25	2.17	2.10	2.04
120	3.92	3.07	2.68	2.45	2.29	2.17	2.09	2.02	1.96
∞	3.84	3.00	2.60	2.37	2.21	2.10	2.01	1.94	1.88

Table A.5—*Continued*

α = .05

10	12	15	20	24	30	40	60	120	∞
241.9	243.9	245.9	248.0	249.1	250.1	251.1	252.2	253.3	254.3
19.40	19.41	19.43	19.45	19.45	19.46	19.47	19.48	19.49	19.50
8.79	8.74	8.70	8.66	8.64	8.62	8.59	8.57	8.55	8.53
5.96	5.91	5.86	5.80	5.77	5.75	5.72	5.69	5.66	5.63
4.74	4.68	4.62	4.56	4.53	4.50	4.46	4.43	4.40	4.36
4.06	4.00	3.94	3.87	3.84	3.81	3.77	3.74	3.70	3.67
3.64	3.57	3.51	3.44	3.41	3.38	3.34	3.30	3.27	3.23
3.35	3.28	3.22	3.15	3.12	3.08	3.04	3.01	2.97	2.93
3.14	3.07	3.01	2.94	2.90	2.86	2.83	2.79	2.75	2.71
2.98	2.91	2.85	2.77	2.74	2.70	2.66	2.62	2.58	2.54
2.85	2.79	2.72	2.65	2.61	2.57	2.53	2.49	2.45	2.40
2.75	2.69	2.62	2.54	2.51	2.47	2.43	2.38	2.34	2.30
2.67	2.60	2.53	2.46	2.42	2.38	2.34	2.30	2.25	2.21
2.60	2.53	2.46	2.39	2.35	2.31	2.27	2.22	2.18	2.13
2.54	2.48	2.40	2.33	2.29	2.25	2.20	2.16	2.11	2.07
2.49	2.42	2.35	2.28	2.24	2.19	2.15	2.11	2.06	2.01
2.45	2.38	2.31	2.23	2.19	2.15	2.10	2.06	2.01	1.96
2.41	2.34	2.27	2.19	2.15	2.11	2.06	2.02	1.97	1.92
2.38	2.31	2.23	2.16	2.11	2.07	2.03	1.98	1.93	1.88
2.35	2.28	2.20	2.12	2.08	2.04	1.99	1.95	1.90	1.84
2.32	2.25	2.18	2.10	2.05	2.01	1.96	1.92	1.87	1.81
2.30	2.23	2.15	2.07	2.03	1.98	1.94	1.89	1.84	1.78
2.27	2.20	2.13	2.05	2.01	1.96	1.91	1.86	1.81	1.76
2.25	2.18	2.11	2.03	1.98	1.94	1.89	1.84	1.79	1.73
2.24	2.16	2.09	2.01	1.96	1.92	1.87	1.82	1.77	1.71
2.22	2.15	2.07	1.99	1.95	1.90	1.85	1.80	1.75	1.69
2.20	2.13	2.06	1.97	1.93	1.88	1.84	1.79	1.73	1.67
2.19	2.12	2.04	1.96	1.91	1.87	1.82	1.77	1.71	1.65
2.18	2.10	2.03	1.94	1.90	1.85	1.81	1.75	1.70	1.64
2.16	2.09	2.01	1.93	1.89	1.84	1.79	1.74	1.68	1.62
2.08	2.00	1.92	1.84	1.79	1.74	1.69	1.64	1.58	1.51
1.99	1.92	1.84	1.75	1.70	1.65	1.59	1.53	1.47	1.39
1.91	1.83	1.75	1.66	1.61	1.55	1.50	1.43	1.35	1.25
1.83	1.75	1.67	1.57	1.52	1.46	1.39	1.32	1.22	1.00

Table A.5—*Continued*

α = .01

df	1	2	3	4	5	6	7	8	9
1	4052.	4999.5	5403.	5625.	5764.	5859.	5928.	5982.	6022.
2	98.50	99.00	99.17	99.25	99.30	99.33	99.36	99.37	99.39
3	34.12	30.82	29.46	28.71	28.24	27.91	27.67	27.49	27.35
4	21.20	18.00	16.69	15.98	15.52	15.21	14.98	14.80	14.66
5	16.26	13.27	12.06	11.39	10.97	10.67	10.46	10.29	10.16
6	13.75	10.92	9.78	9.15	8.75	8.47	8.26	8.10	7.98
7	12.25	9.55	8.45	7.85	7.46	7.19	6.99	6.84	6.72
8	11.26	8.65	7.59	7.01	6.63	6.37	6.18	6.03	5.91
9	10.56	8.02	6.99	6.42	6.06	5.80	5.61	5.47	5.35
10	10.04	7.56	6.55	5.99	5.64	5.39	5.20	5.06	4.94
11	9.65	7.21	6.22	5.67	5.32	5.07	4.89	4.74	4.63
12	9.33	6.93	5.95	5.41	5.06	4.82	4.64	4.50	4.39
13	9.07	6.70	5.74	5.21	4.86	4.62	4.44	4.30	4.19
14	8.86	6.51	5.56	5.04	4.69	4.46	4.28	4.14	4.03
15	8.68	6.36	5.42	4.89	4.56	4.32	4.14	4.00	3.89
16	8.53	6.23	5.29	4.77	4.44	4.20	4.03	3.89	3.78
17	8.40	6.11	5.18	4.67	4.34	4.10	3.93	3.79	3.68
18	8.29	6.01	5.09	4.58	4.25	4.01	3.84	3.71	3.60
19	8.18	5.93	5.01	4.50	4.17	3.94	3.77	3.63	3.52
20	8.10	5.85	4.94	4.43	4.10	3.87	3.70	3.56	3.46
21	8.02	5.78	4.87	4.37	4.04	3.81	3.64	3.51	3.40
22	7.95	5.72	4.82	4.31	3.99	3.76	3.59	3.45	3.35
23	7.88	5.66	4.76	4.26	3.94	3.71	3.54	3.41	3.30
24	7.82	5.61	4.72	4.22	3.90	3.67	3.50	3.36	3.26
25	7.77	5.57	4.68	4.18	3.85	3.63	3.46	3.32	3.22
26	7.72	5.53	4.64	4.14	3.82	3.59	3.42	3.29	3.18
27	7.68	5.49	4.60	4.11	3.78	3.56	3.39	3.26	3.15
28	7.64	5.45	4.57	4.07	3.75	3.53	3.36	3.23	3.12
29	7.60	5.42	4.54	4.04	3.73	3.50	3.33	3.20	3.09
30	7.56	5.39	4.51	4.02	3.70	3.47	3.30	3.17	3.07
40	7.31	5.18	4.31	3.83	3.51	3.29	3.12	2.99	2.89
60	7.08	4.98	4.13	3.65	3.34	3.12	2.95	2.82	2.72
120	6.85	4.79	3.95	3.48	3.17	2.96	2.79	2.66	2.56
∞	6.63	4.61	3.78	3.32	3.02	2.80	2.64	2.51	2.41

Table A.5—*Continued*

α = .01

10	12	15	20	24	30	40	60	120	∞
6056.	6106.	6157.	6209.	6235.	6261.	6287.	6313.	6339.	6366.
99.40	99.42	99.43	99.45	99.46	99.47	99.47	99.48	99.49	99.50
27.23	27.05	26.87	26.69	26.60	26.50	26.41	26.32	26.22	26.13
14.55	14.37	14.20	14.02	13.93	13.84	13.75	13.65	13.56	13.46
10.05	9.89	9.72	9.55	9.47	9.38	9.29	9.20	9.11	9.02
7.87	7.72	7.56	7.40	7.31	7.23	7.14	7.06	6.97	6.88
6.62	6.47	6.31	6.16	6.07	5.99	5.91	5.82	5.74	5.65
5.81	5.67	5.52	5.36	5.28	5.20	5.12	5.03	4.95	4.86
5.26	5.11	4.96	4.81	4.73	4.65	4.57	4.48	4.40	4.31
4.85	4.71	4.56	4.41	4.33	4.25	4.17	4.08	4.00	3.91
4.54	4.40	4.25	4.10	4.02	3.94	3.86	3.78	3.69	3.60
4.30	4.16	4.01	3.86	3.78	3.70	3.62	3.54	3.45	3.36
4.10	3.96	3.82	3.66	3.59	3.51	3.43	3.34	3.25	3.17
3,94	3.80	3.66	3.51	3.43	3.35	3.27	3.18	3.09	3.00
3.80	3.67	3.52	3.37	3.29	3.21	3.13	3.05	2.96	2.87
3.69	3.55	3.41	3.26	3.18	3.10	3.02	2.93	2.84	2.75
3.59	3.46	3.31	3.16	3.08	3.00	2.92	2.83	2.75	2.65
3.51	3.37	3.23	3.08	3.00	2.92	2.84	2.75	2.66	2.57
3.43	3.30	3.15	3.00	2.92	2.84	2.76	2.67	2.58	2.49
3.37	3.23	3.09	2.94	2.86	2.78	2.69	2.61	2.52	2.42
3.31	3.17	3.03	2.88	2.80	2.72	2.64	2.55	2.46	2.36
3.26	3.12	2.98	2.83	2.75	2.67	2.58	2.50	2.40	2.31
3.21	3.07	2.93	2.78	2.70	2.62	2.54	2.45	2.35	2.26
3.17	3.03	2.89	2.74	2.66	2.58	2.49	2.40	2.31	2.21
3.13	2.99	2.85	2.70	2.62	2.54	2.45	2.36	2.27	2.17
3.09	2.96	2.81	2.66	2.58	2.50	2.42	2.33	2.23	2.13
3.06	2.93	2.78	2.63	2.55	2.47	2.38	2.29	2.20	2.10
3.03	2.90	2.75	2.60	2.52	2.44	2.35	2.26	2.17	2.06
3.00	2.87	2.73	2.57	2.49	2.41	2.33	2.23	2.14	2.03
2.98	2.84	2.70	2.55	2.47	2.39	2.30	2.21	2.11	2.01
2.80	2.66	2.52	2.37	2.29	2.20	2.11	2.02	1.92	1.80
2.63	2.50	2.35	2.20	2.12	2.03	1.94	1.84	1.73	1.60
2.47	2.34	2.19	2.03	1.95	1.86	1.76	1.66	1.53	1.38
2.32	2.18	2.04	1.88	1.79	1.70	1.59	1.47	1.32	1.00

INDEX

About the Author

WAYNE C. CURTIS is Dean of the Sorrell School of Business and Commerce at Troy State University in Troy, Alabama. His articles on law, economics, and statistics have been published in *Jurimetrics Journal, The Atlantic Economic Journal, The Journal of Behavioral Economics, The Alabama Lawyer, Review of Business and Economic Research,* and *American Journal of Agricultural Economics.*

www.ingramcontent.com/pod-product-compliance
Ingram Content Group UK Ltd.
Pitfield, Milton Keynes, MK11 3LW, UK
UKHW020807290926
15470UKWH00003B/998

9 780899 300337